Creative Brickwork

BRIDGWA

This book must
below.

BRIDGWATER COLLEGE

B0073072

CANCELLED

Creative Brickwork

Terry Knight

On behalf of the Brick Development Association

ARNOLD

A member of the Hodder Headline Group
LONDON • SYDNEY • AUCKLAND

BRIDGWATER COLLEGE LRC

Greenhead	16.06.98
B0073072	£15.99
693.21 KNI	

First published in Great Britain 1997 by Arnold
a member of the Hodder Headline Group,
338 Euston Road, London NW1 3BH

© 1997 The Brick Development Association

All rights reserved. No part of this publication may be reproduced
or transmitted in any form or by any means, electronically or mechanically,
including photocopying, recording or any information storage or retrieval
system, without either prior permission in writing from the publisher or a
licence permitting restricted copying. In the United Kingdom such licences
are issued by the Copyright Licensing Agency: 90 Tottenham Court Road,
London W1P 9HE.

Whilst the advice and information in this book is believed to be true and
accurate at the date of going to press, neither the author nor the publisher
can accept any legal responsibility or liability for any errors or omissions
that may be made.

British Library Cataloguing in Publication Data
A catalogue record for this book is available from the British Library

ISBN: 0 340 67643 4

Typeset in 10.5/11.5 Plantin by J&L Composition Ltd, Filey, North Yorkshire
Printed and bound in Great Britain by St Edmundsbury Press, Bury St Edmunds, Suffolk
and J.W. Arrowsmith Ltd., Bristol

Contents

Acknowledgements

Covering so many aspects of a subject in one book clearly needs the expert guidance of specialists. Only the author can fully appreciate the patience of the many individuals from the Brick Development Association who commented on early drafts. That they were so ably supported by members of BDA's staff came as no surprise to a former colleague. But it would not be appropriate to mention individuals in such a corporate team.

It is only right that I should thank Terence P Smith BA, MA, MLitt, Chairman of the British Brick Society and David Kennett BA MSc, Editor of *Information* published by the Society. They have both achieved much in recent years in encouraging and advancing the serious study of the history of bricks and brickwork and I am grateful for their firm but kindly guidance in my early attempt to provide a brief historical introduction.

My thanks also, for their special guidance, to Bob Baldwin PPGB, brickwork consultant and Mick Pearce PPGB, Associate CIOB who have for many years through the Guild of Bricklayers supported the Brick Development Association and its members in promoting high quality brickwork; and to Gerard Lynch a master bricklayer, lecturer, consultant and author.

Photographs and illustrations have been supplied by Michael Hammett, Martine Hamilton Knight, Terry Knight and Frank Walter.

Although I have taken note of all comments I take full responsibility for the final content and any errors.

Terry Knight
1996

Introduction

Looking at brickwork

Travelling through town and country we may wonder at the variety of brick colours and textures with which the United Kingdom is favoured, and appreciate that the wide variety of clays stems from the complex geology of this country (**Plate 1**). Any passing brickmaker would confirm that much of the variety stems from the ancient and modern art and science of brickmaking.

At first the very profusion of brickwork may well blunt our appreciation of this abundant, indigenous and versatile material. Yet experience suggests that the more we look, learn and understand, the more we appreciate and enjoy our surroundings. Finally, we may appreciate the talents of those that designed and built such a wealth of brick architecture.

About this book

This book identifies design and building decisions and actions that determine the appearance of both new brickwork and that subject to conservation.

It is intended to aid practitioners, students and trainees to exercise their own skills but also, through a better understanding, to appreciate the skills of others.

It will also help those whose job is to commission, brief, approve or regulate building work to understand the potential of brickwork in our environment. These may include industrial and commercial clients, school governors, tenant groups and politicians.

1 A brief history of bricks in the UK

1.1 Roman bricks, their use and re-use

From AD 43 for about 400 years the Romans in Britain made and built with fired-clay bricks of many sizes. They were typically about 18 × 12 × 1 to 1½ inches thick and were often used as lacing or bonding courses. At a Roman fort in Burgh Castle, Suffolk, rubble walls faced with flint contain three courses at vertical intervals of a little over one metre (**Fig. 1.1**). Other bricks were square, from 9 × 9 inches to 24 × 24 inches and were used to build 'pilae' (supports) for hypocaust heating systems. Rectangular bricks between 9 × 7½ inches to 15 × 10½ inches were used to turn arches. Brick was a utilitarian, mainly structural material often hidden from sight by rendering or other facing materials so that appearance hardly mattered.

The end of the Roman occupation was marked by the formal severence in AD 410 when a document known as the Rescript of Honorius told its recipients to 'look to their own defences'. Brick-making virtually ceased for 600 to 700 years after the Roman legions left.

The Anglo-Saxons took the opportunity to move in and govern much of Britain. Coming from parts of northern Europe lacking good building stone, but

Fig. 1.1 Brick lacing courses in rubble wall of a Roman Fort, AD 30.

abundant in forests, they built mainly in wood, although, following the gradual conversion to Christianity after the arrival of St Augustine of Canterbury in AD 597, they built a number of churches mainly of rubble with ashlar masonry at the quoins and reveals to door and window openings.

As the Roman buildings fell into disuse and ruin, their bricks were recycled for use in new buildings, although it has been suggested that at Brixworth Church in Northants, believed to be 9th century, the bricks used to build arches may in fact be Anglo-Saxon. Thermoluminescence tests have been inconclusive.

Anglo-Saxon use of bricks was mostly confined to the construction of quoins and door and window openings which were then rendered. The decorative use of Roman bricks in the tower of Holy Trinity Church, Colchester, was an exception.

1.2 Slow development from 1066

Shortly after the Conquest there began a slow but steady development in the use of brickwork which preceded a great upsurge of brick building around Tudor times.

Just 10 or 12 years after their conquest of Britain the Normans recycled Roman bricks for use at Colchester Castle. They were used to build quoins and as 'lacing' courses to bond the flint rubble structure as can be seen in the walls at Verulamium (St Albans). During the 12th century they used Roman bricks to build the upper stages of the tower of St Albans Abbey (**Plate 2**). As the original rendering to the brickwork no longer exists the red brickwork can be seen, and very handsome it is too. But, generally, stone was the Norman choice for masonry construction.

An early example of the revival of brickmaking and brickwork in England, is St Mary's church Polstead, Suffolk,

c.1160. It was a prestigious building, much larger than most village churches. The bricks are 10 to 11 × 5 to 7 × 1¾ inches. It has been suggested that the church can be seen as a 12th century precursor of the grand 15th century churches of Suffolk.

Brickwork was used in the building of Little Coggeshall Abbey and the nearby Chapel of St Nicholas in Essex, 1185 and 1220 respectively. The Abbey includes early medieval warm red bricks (12 × 5 to 6 × 1¾ inches)(**Fig. 1.2**) The main walling of the chapel is flint rubble with bricks confined to quoins and dressings to window openings, and there are some moulded brick details.

Little Wenham Hall near Ipswich, Suffolk, 1270 to after 1287, is thought to be the first domestic building built largely in brick. The bricks are described as being of a true Flemish or Low Countries type, mostly cream and greenish yellow with occasional pinks and reds. They are 8¼ to 9 × 4¼ to 4½ × 2 inches thick, built four courses to 10 inches. It is said to be one of the earliest buildings in this country in which the brickwork seems familiar to us today. The bricks, being variable in length, are not regularly bonded. The bulk of the structure is in brick but the piers and dressings to the windows which include carved mullions and tracery are in stone. The use of brick was to become more frequent in the 14th century.

In Kingston upon Hull, a town which did much trade with northern Germany

Fig. 1.2 Bricks at Little Coggeshall Abbey, Essex c.1200.

and the Low Countries, the Corporation operated a town brickworks from 1303. The town was built largely of brick including the encircling walls, gate-houses and river embankment. Little remains visible except for Holy Trinity Church built in red brick with stone dressings, and excavated ruins of parts of the enclosing wall.

Other early examples are the gate-house of Thornton Abbey, Lincolnshire, c.1382 built of red brick contrasted with Ashlar, and the late 14th century Cow Tower, part of the defences of the city of Norwich. The walls of the tower have a flint core entirely faced with brick.

1.3 Increasing development – 13th to 15th centuries

a) Favourable conditions

A steady growth in the manufacture and use of bricks in the eastern counties of England and the Thames Valley contin-ued because a number of favourable con-ditions either existed or developed together.

- There were numerous deposits of soft alluvial clays close to the surface which could be dug and prepared by hand, and there was poor quality wood close by for firing.
- In general a lack of good quality ash-lar stone favoured the growth of brick-work, as it had in northern continental Europe. But it is interesting that Lin-colnshire, where there was much good stone, has many examples of early brickwork.
- The slow revival of brickmaking that had begun in the 12th century was sustained around the middle of the 13th century by immigrant brickmak-ers from northern Europe, and by Flemish refugees.

- A growing trade, largely in wool with northern continental Europe and the Low Countries, brought about the ex-change of architectural influences be-tween English merchants and those of the Hanseatic League, many of whom established bases in English ports.

b) Brick becomes a prestige material

Patronage and money flowed from the Kings of England and those loyal sup-porters whom they made wealthy by the gifts of land. During the so-called Hun-dred Years War with France, which con-tinued on and off from 1333 to about 1453, many nobles and knights returned home rich enough to build grandly.

As a result, throughout the 15th cen-tury, brick became a prestige material for large buildings such as castles and col-leges. They were built by a small group of influential and very wealthy people who were close to King Henry V and King Henry VI who between them reigned from 1413 to 1461. The use of brickwork for buildings such as the examples noted in d) below proclaimed the wealth and high social standing of their owners. They were rich enough not only to use brick but to exploit its versatility by enhancing buildings with patterns, elaborate chimneys, corbelling and moulded brickwork.

c) Exploiting the versatility of bricks

Patterns in brickwork using bricks that were darker than the rest of the brickwork are thought to have been introduced from France in the 15th cen-tury. The dark, grey-blue or purple bricks that were generally used were headers that had been overburnt or pos-sibly vitrified. The headers were not uniform in colour, as to some extent they were the kilnburners' failures. Some hea-ders were so light in colour, the pattern tended to fade giving a more subtle effect than the crisp appearance obtained today when using comparatively dark, uniform modern bricks.

Diapers became increasingly popular during the mid 15th century and continued to develop during the early Tudor period, 1480s to 1520s, into very elaborate patterns as in the Bishop of Ely's Palace, Hatfield, Hertfordshire c.1480. The commonly used English Bond was usually modified to allow the use of extra headers which generally formed the pattern. English Cross Bond was also used at times. By 1520 there were many examples of walls decorated with diaper pattern.

Brick chimneys were often preferred for houses otherwise built from stone. Presumably brick was found more suitable than stone for building the intricacies of flues and was further commended for having been fired at high temperatures. But there may have been another reason.

The appearance of chimneys in the prestigious buildings of the 15th century signalled that the owner had a house with many fireplaces. But the very nature of brickwork also gave an opportunity for decorative intricacy and flamboyance. In fact some chimneys were fakes, built only for their visual impact.

The most flamboyant examples are probably those at Hampton Court which were built to function. That most of them are excellent Victorian replicas and replacements reminds us that chimneys are amongst the most exposed of building elements and require special care in the specification of bricks and mortar and detailed design.

Corbelling was often employed in medieval castles and fortified manor houses for the construction of machicolations, which provided vertical holes under the battlements down which hot liquids or solid missiles could be dropped on an enemy below. When machicolations were no longer needed for defence, new buildings often incorporated imitations for the sake of appearance, an example of a human inclination to retain at least the likeness of familiar artifacts long after the original need has ceased. Corbelling was also used to build out an eaves course to simulate a classical entablature, to allow a gable to oversail at an upper storey, and to support a right angled corner over a splayed corner below or a splay over an internal angle. It may be that it is because corbelling is so characteristic of brickwork that we find such features attractive.

Corbel tables. A corbel table is a continuous range of corbels, supporting a projecting parapet or machicolations. It was a feature that lent itself to many interpretations using not only normal bricks but also moulded and carved bricks.

Moulded bricks were used as dressings around arches, windows and for other special uses contrasting with the main brickwork. Window openings often contained moulded brick tracery although stone was also used in otherwise brick buildings. In the Essex–Hertfordshire area wholly brick tracery was introduced for church windows during the early Tudor period.

d) Some examples

Gatehouses, probably the most substantial part of many castles and fortified manor houses, have often survived the other parts. Gatehouses continued to be built during the 15th century although less and less to meet a defensive need. They did, however, continue to signal the importance of the individual or institution that owned it. Prominent examples are the Cambridge colleges: namely Queens' College, 1448; Jesus College, c.1500 and St John's College, 1511 onwards; as well as Layer Marney, Essex, c.1525. The practice of building such gatehouses seems to have ceased with few exceptions by the beginning of Elizabeth's reign in 1558.

Tattershall Castle in Lincolnshire, for Ralph, Lord Cromwell who was in France during most of the reign of Henry V, 1413 to 1422, was finished about 1440 (**Fig. 1.3**). To many at the time the tower must have looked foreign but less so to us, accustomed as we are to many other English gatehouses and towers. The red/brown bricks, about 2 inches thick, were made 10 miles away at Edlington Moor. Its features include diaper patterns, battlements, stone corbelled machicolations, stone dressings to windows, string courses and copings. It was restored during the 20th century.

Herstmonceux Castle in Sussex, c.1441, was built by Sir Roger de Fynes who fought at Agincourt in 1415 and elsewhere. The bricks are a light pinkish colour laid in English Bond with paler bricks used to form a delicate diaper pattern. The castle was one of the most

Fig. 1.3 Tattershall Castle, Lincolnshire, c.1440.

westerly major brick buildings in England at the time.

At **Rye House**, Stanstead Abbotts in Hertfordshire, the gatehouse, 1440, was built from rough, rich red bricks with some pebbles evident. Light, rather than darker, contrasting blue headers provide 'soft' diaper patterns. There are two corbelled-out oriel windows on the first floor, moulded brick labels over ground floor windows, and under the oriel windows and over the entrance door are corbel tables with cusped trefoil heads.

Eton College, Buckinghamshire, was built during the 1440s for which millions of bricks were made at Slough. The brick buildings of this period that remain include 'Founders Tower' and most of College Cloister.

Queens' College Cambridge, founded and built in the 1450s, is one of the finest examples of brickwork in Cambridge and has been altered less than other brick colleges in the city.

Hampton Court Palace, the epitome of Tudor architecture, owes its existence to Thomas Wolsey who amassed so much wealth as Cardinal and Lord Chancellor to Henry VIII he could afford to build a magnificent country home by the Thames on a site leased in 1514. Some 15 years later, following Wolsey's fall from favour, power and wealth, it became Henry's Royal Palace (**Plate 3**).

The brickwork relies for effect mainly on the bricks' damson, plum and russet colours but also on their rough texture. The whole is complemented by the stone dressings, generally at window and door openings, a feature that is to be seen in many earlier brick buildings built in the 15th century (**Plate 4**). The only purpose-made moulded bricks are those used in the many elaborate differing chimneys (**Plate 5**)(**Fig. 1.4**). There is diapering in some of the courts. Brick, a relatively expensive building material, was used because it was a highly fashionable and prestigious material and only a few very

Fig. 1.4 *19th century restoration of chimneys at Hampton Court, 1520. A prime example of the early use of purpose-made moulded bricks to build elaborate chimneys which displayed the wealth and importance of the owner.*

1.4 Reaching other places and other people with other styles – from mid 16th century

a) Brickwork usage spreads

About mid 16th century, brick began to spread beyond the eastern counties of England and the Thames Valley where it began, but tended to 'jump' the limestone belt which runs in a broad irregular band in an arc from Somerset to Durham. By the end of the 18th century brick building could be found in all English counties except the very South-West and the far North-East. This is not to deny that isolated instances of the use of brickwork in these 'new' areas had already occurred. For example, Yorkshire's East Riding had a tradition of brickwork going back to the 14th century.

Bricks and brickwork were about to adapt to meet the needs of changing architectural style. More buildings were to have classical details either carved in stone, or built from moulded bricks which were sometimes rendered to resemble stone. At the same time the use of bricks was spreading across the country and down the social scale.

b) The picturesque gives way to classical order

The picturesque nature of Tudor architecture emanated from broken outlines of turrets, battlements and elaborate chimneys, and from mullioned and transomed windows, as well as the irregularity and rugged texture of the bricks. This gradually gave way to the regularity of classical columns, entablatures and pediments, complemented by more regular and smooth bricks as well as advances in bricklaying techniques. Late

rich people in the Court circle could afford it.

Bricks were still being made on or close to the building sites. Firing in both clamps and kilns produced a great variety of colour, even within one firing, and generally they were rough in texture and imprecise in shape and dimensions. Quite a number of bricks had flared headers of a different colour from the stretcher faces and were used to create characteristic Tudor patterns.

Although Tudor bricks are often thought of as being a brilliant red, there were many darker and lighter shades of red, browns, mauves, blues and even greys. They were generally built in at random, as they came to hand, giving the typical Tudor appearance.

in the century, sliding sash windows began to take over from mullioned and transomed windows.

Diapers too were waning; presumably they began to look old fashioned. An exceptional example was the use of regular diaper patterns on a classical façade at Sudbury Hall built in Derbyshire in the early 17th century.

Dressings at quoins and window openings were more often used during the second quarter of the century. Bright red bricks often contrasted with those in the main wall. Some further refinement was also introduced at this time by the use of gauged brickwork.

c) Styles in transition

Mapledurham House, Oxon, c.1585, has been instanced as a transitional building having an early Tudor look with its gabled sky-line and prominent chimneys. But the large windows and the symmetrical façade are entirely Elizabethan. A further example is Cobham Hall, Kent, c.1584, which instead of gables has a straight parapet interrupted by conspicuous chimneys and behind which the roof is prominent. It has a classical entry but the elevation is not quite symmetrical.

Moyns Park, a moderate-size house in Essex, c.1580, had stone quoins and dressings used sparingly. It was almost a prototype for some of the larger houses and palaces that were to follow in the early 17th century. One such was Charlton House in Kent, 1607, which instead of the steeply pitched roof and gables at Moyns Park had a low pitched roof, mainly hidden behind a pierced parapet. The ultimate in this development was probably Hatfield House, Hertfordshire, 1608–13.

Kew Palace, 1631 (**Plate 6**) (**Fig 1.5**), was one of the first buildings to use Flemish rather than English Bond which thereafter went out of fashion. It was built by Samuel Foutrey, of Dutch descent, which is probably why, together

Fig. 1.5 Gauged and carved Corinthian capitals and key block over windows at Kew Palace, c.1631.

with the form of the gables, it was known for many years as the Dutch House. The theme was followed by many types of building including richly decorated brickwork at Broome Park, Kent, 1635–38, and Cromwell House Highgate, London.

The serene quality of Flemish Bond, compared with the more linear pattern of English Bond, would seem to have made it a natural and popular choice for the modern architecture of the day; but it has to be said that in The Netherlands, with its no less classical architecture, the use of English Bond, English Cross Bond or Dutch Bond continued unabated.

d) Increasing formality

During the 17th century, architecture became more formal, restrained and even severe. It was greatly influenced by Inigo Jones who, following his return from Italy in 1615, was inspired by the formal rules of Palladian architecture. At the time the starkness of his work was not generally popular outside court circles but it did influence those whose wealth enabled them to build fashionably.

During the reign of Charles I, 1625–49, some four-storey housing built in Great Queen Street, London was later described as 'the first regular street in London'. John Summerson commented, 'They laid down the canon of street design which put an end to gabled individualism, and provided a discipline for London's streets which was accepted for more than two hundred years.'

Many developments used brick dressings instead of stone, as at Balls Park, Hertfordshire, c.1640. At Tyttenhanger Park, St Albans, 1654, quoins and window dressings are in a darker brick while the mouldings of the pediments and window dressings are of cut or carved brick.

e) The influence of Sir Christopher Wren

Bricks have often been used where their contrasting colour and scale complements and enhances the stone structure. This would often appear to be an aesthetic choice rather than because brickwork was cheaper than stone. An example is in the Middle Temple Gateway in London by Sir Christopher Wren where small, soft, rubbed bricks, 6×2 inches, with very fine joints nicely set off the stone pilasters.

Wren is remembered for his great projects such as St Paul's Cathedral, 1675–1710, many London churches and Greenwich Hospital, 1696–1715. But it was the new Hampton Court Palace begun in 1689, the year following the accession of William III and Mary II, that was greatly to influence the use of brickwork. Red brick set against white stone or paintwork as part of a modest restrained design is sometimes referred to as 'Wren' architecture. Wren himself used it in a few private houses whose

serenity derives from their simplicity and good proportions, not from expensive elaboration. At the church of St Benet, Paul's Wharf the delicately carved stone swags are set off by the red bricks.

The style which is so often associated with the reign of Queen Anne, 1702–14, was emulated by many others.

1.5 The Georgian era 1714–1837

The use of brickwork was to change in two important ways during the 18th century. With growing prosperity and the pause in the war with France following the Treaty of Utrecht in 1713, building increased and architectural style changed as an aftermath of Inigo Jones' influence. John Summerson commented that:

> Yet taste has played as great a part as money, for taste has always formed, and still forms, a vital factor in the economics of building. It gives the psychological *élan* to the building market corresponding to the speculator's urge to enrich himself. Taste and wealth – these two are the basic things in Georgian London . . .

Much of the building in London during the rest of the 18th century was by speculators conforming to the generally established 'Georgian' style, still owing much to Palladian architecture but mostly in brick rather than painted stucco. Much of it was to provide town houses for country gentlemen and continued through to the 19th century (**Plate 7**) (**Fig. 1.6**). By about 1720 new brickwork in London was of pale, yellowish brown London Stocks rather than the red of Queen Anne's reign, 1702–14. Elsewhere reddish brown bricks with yellow tinges often referred to as grey were used. The truly grey bricks were often used for the better work. For example, at Holkham Hall, Norfolk, 1734–36, by William Kent, the main walls and the

Fig. 1.6 House in Bedford Square, London, c.1780.

rusticated base are built in light grey bricks.

Stock bricks from dark red to a dull pink were beginning to be used for dressings to enhance the brickwork or at times just for arches. Often brickwork dressings were of special bricks called cutting bricks, rubbing bricks or 'rubbers' (**Fig. 1.7**). They were usually laid in putty mortar to achieve fine joints so that the dressings were in complete contrast to the normal brickwork and stood out as did stone dressings. In general, brickwork joints became narrower and tuck pointing (described in Section 2.2.3 h) was developed to imitate fine-jointed gauged brickwork. Often the brickwork was blackened and the mortar joints finished with tuck pointing in a contrasting white lime putty.

To summarize, there was a reaction against red brick although it continued to be used by some 'Palladians' and also for less important façades of buildings. Yellow-brown was much used in London, 'white' and silver-grey were often used in the Thames Valley. Flemish Bond was practically universal but with some Header Bond especially in southern counties.

Many timber-framed buildings were faced in brick either to follow fashion or

Fig. 1.7(a) The features are in red gauged brickwork, the main walling is in silver-grey bricks.

Fig. 1.7(b) A close comparison of the two types of brickwork.

to improve fire resistance. This was sometimes done, mainly in the South East, using mathematical tiles that can be considered as a special variation of plain clay roof tiles where the appearance of brickwork was required without its weight.

1.6 19th century

a) A vast increase in brick usage

During the 19th century there was a vast increase in brick manufacture to meet the growing demand for housing, churches, public buildings, factories and ware-houses. The railways created a major demand for bricks between 1830 and 1850.

Many civil engineering projects re-quired strong and durable bricks such as Accrington Reds, Southwaters and Staffordshire Blues, although the last, in particular, were often used for purely decorative purposes in polychromatic brickwork.

Demand was met by making use of colliery shales and later the Lower Oxford clay mainly in Bedfordshire. Mechaniza-tion led to the pressing of bricks which produced a more precise and regular form, sharp arrises, more uniform colour and smooth texture, all of which gener-ally the Victorians seemed to favour. There was, however, still some demand for thin bricks for the vernacular revival of the arts and crafts style and moulded bricks for the buildings of the Gothic revival.

In 1880 the Lower Oxford clay began to be used for making Flettons to meet the enormous demand for building, par-ticularly housing, during the last years of the century. Brickmaking was helped along by the advent of the steam shovel and dragline excavators and the multi-chambered, continuous burning Hoff-mann kilns invented in Germany in 1859.

The mechanization of manufacture did not mean the end of small local works using more primitive methods, and for whose products there remained a considerable demand particularly for the less regular bricks which many admired in older brickwork.

b) Changing architectural use

Flemish Bond continued to be popular but English Bond enjoyed a rebirth partly for its perceived greater strength for engineering projects and partly for Gothic revival buildings and utilitarian brickwork in the North. There was some use of Header Bond in the North West and Manchester. At the end of the cen-tury cement began to supplant lime in mortars, and the use of ash for black mortar became popular in industrial areas and also in London. Tuck pointing continued but became outmoded by the end of the century. Even quite humble housing was embellished with dentil or dog tooth features or bands of coloured bricks.

One characteristic image of the Victorian age is polychromatic brick-work making use of the wide variety of brick colours available. Church archi-tects prized bricks for both the massive-ness and the colour they could impart.

One of the earliest was James Wild's Christ Church, Streatham, Lon-don, 1840, which uses colour sparingly for such a massive structure. By contrast William Butterfield's All Saints Church, Margaret Street, Marylebone, 1849–53, built at a time when colour was gaining favour, did not meet with everyone's approval and some described it as coarse while others wondered whether Butter-field had a 'deliberate preference for ugli-ness'. To this day his work arouses opposite passions in different people. His Magnum Opus was Keble College, Oxford (**Plate 8**) (**Fig. 1.8**), begun in 1868–82, which has vivid polychromy in brick and stone horizontal stripes with architecture but was well suited to urban housing and was understood by many people, perhaps because it so often used familiar materials, like brickwork, in a variety of straightforward ways.

The first university lectures for women having started in 1870, the Old Hall of Newnham College Cambridge by Basil Champneys, was opened in 1875 in what Pevsner referred to as the Dutch red brick style in contrast with the Gothic and Tudor of the men's colleges. It is notable for its fine brick detailing (**Fig. 1.9**).

c) A monumental end to the century

The architect Richard Norman Shaw declared of Westminster Cathedral in London, that 'beyond doubt, it is the finest church built for centuries'. Others thought it too exotic and 'un-English'. Certainly the structural brick form is magnificent, the more surprising as the architect J F Bentley was convinced that Gothic was the only true style for a church while his client Cardinal Vaughan was adamant that it should not compete with the nearby 'real' medieval Gothic of Westminster Abbey. Following a study tour abroad Bentley produced the Byzantine style design which enabled it to be built quickly and cheaply in structural brickwork, avoiding the costly carved ornament of Gothic architecture. Millions of common Fletton bricks were used in the structure. Externally, red hand-made bricks are laid in an English Garden-Wall Bond setting off the white Portland stone dressings and echoing the character of the Queen Anne revival common in London at that time. The internal brickwork is in English Bond and is extensively covered in marble and mosaic except at the higher levels where the cladding remains incomplete and the magnificent, monumentality of the brick structure is still evident.

Fig. 1.8 Keble College, Oxford, 1868–82.

striking zig-zag and diaper patterned brickwork above. The fashion was probably out of date by the time it was completed. Stefan Muthesius commenting on St Bartholomew Church, Brighton, 1874, says it is 'Perhaps the most impressive of all "town churches" although its polychromy must have been considered rather old fashioned at the time.'

Massiveness, variety of form and bold polychromatic decoration were superseded by a simpler form exemplified by Red House designed in 1859 by Philip Webb for William Morris. Richard Norman Shaw also looked back to the simple brick dwellings based on classical architecture nearly 200 years earlier and which had been dubbed 'Queen Anne style', in which features such as pilasters and white-painted windows were set off by the colour of the brickwork. It was a homely style which might never be great

1.7 20th century

Fig. 1.9 Newnham College, Cambridge, 1875–1900.

a) Between the two world wars

Most new housing after the 1914–18 war continued to be built with solid one-brick thick external walls.

Some of the most attractive housing layouts which contained much simple competent brickwork were local authority developments some probably influenced by Letchworth Garden City, begun during Edward VII's reign and by Welwyn Garden City in 1919. They were generally brick-built suburban estates of two-storey terrace houses, but in metropolitan areas there were estates of loadbearing brick-built flats generally of no more than five storeys. The brickwork was generally plain but with some restrained decoration.

Speculatively built housing for sale proliferated during the last part of the 1920s and through the 1930s until the Second World War began in 1939. They were mostly two-storey semi-detached and basically brick-built, but with numerous variations to imply individuality and exclusivity. Features included complete or partial rendering, painting or pebble dashing, tile hanging, wany edged hardwood boarding or surface fixed 'mock Tudor' timber framing. The houses were generally well built and met a growing demand for house ownership encouraged by the growth of railways and in particular the extension of the London Underground lines into the suburbs.

Most large buildings, such as office blocks, were of structural steel and later reinforced concrete frames which were often clad in brickwork, many in a style known as Neo-Georgian. Many had pitched roofs behind parapet walls but some had overhanging eaves.

Among the most interesting buildings architecturally were the new London Underground stations, such as those at Osterly, Sudbury Town, Park Royal, Rayners Lane and Arnos Grove.

b) Post Second World War

The introduction of the cavity wall literally changed the face of brickwork to an almost universal Stretcher half-bond.

The desperate post-war need for new housing propagated hundreds of proposals for new non-traditional systems for constructing houses, but only a handful were used. The structures were steel, reinforced concrete or timber frames, *in situ* concrete and in one case a brick-faced precast concrete panel. They were mostly untried in production and use but with the encouragement of government competed with traditional methods of building both low-rise and multi-storey housing.

Very few of the systems are still in use. They were overtaken by 'rationalized traditional' building which aimed to use but develop traditional methods.

c) A revived interest in brickwork

Some of the best local authority housing in the 1970s and 1980s was brick-faced as was new housing built for sale to meet buyers' acknowledged preference.

During this period there was a massive swing back to brickwork for most types of building, probably largely as a reaction against facing concrete which seldom weathered well in the United Kingdom. The rekindled interest in brick was accompanied by a noticeable increase in the use of special shaped bricks, which gave designers the freedom to design in detail the appearance of a building. See Section 2.1.1 g.

Bricks of all colours and textures are now often used to suit the context of the new building as well as the building itself. Recently, multi-rough clay stock bricks complementing stained timber framing were used in two-storey offices in a mature parkland of lawns and trees in Abingdon, Oxfordshire (**Plate 9**). The same architect followed this with a sixth form school on an open, featureless site using pale yellow calcium silicate bricks both internally and externally. Their crisp precision complemented the finely detailed open steel framing to the roof structure (**Plate 10**).

Throughout time a general tendency for one colour to be preferred over another can be discerned. A preference for brown bricks in the early 1970s gave way to one for red bricks in the 1980s, particularly the hard smooth reds similar to engineering bricks. Most recently in the 1990s there seems to be a preference for light buffs and pale yellows, particularly amongst those providing additions to Oxford and Cambridge colleges and probably mindful of the prominence of stone in those places.

The brick industry responded to the increased demand by constructing modern plants equipped with computer managed tunnel kilns capable of making large quantities of bricks to a consistently high quality. Small moving hood and shuttle kilns have also been introduced to meet small, special orders without interrupting the main production line. There are still a number of small brickworks making traditional bricks for which there is a demand for both new and conservation work.

This should surprise no one, as although many building requirements may change, there is still a demand for such a tried, trusted, adaptable and indigenous material that shows every sign of continuing to meet our changing needs as it has done for over a thousand years.

d) The Brick Development Association

From funds provided by the government to compensate brick manufacturers forced to close down during the Second World War the Brick Development Association was set up to assist retraining and re-establishment of the industry on the cessation of hostilities. Later the role of the Association changed to one with an increased concern for the provision of technical information on the design and construction of brickwork and an emphasis on promotion.

The Association is also concerned that bricklayers should be trained not only in the skills of laying bricks but also in an understanding of brick as a material and the reasons behind many of the design decisions and instructions. To this end, and in consultation with, and with the direct cooperation of, brickwork instructors it produced in 1993 a manual called *Achieving Successful Brickwork*. In 1994 this was jointly published with Edward Arnold in a paperback version entitled the *BDA Guide to Successful Brickwork* which is referred to for more detailed information in many sections of this book.

2 Design decisions that affect appearance

The factors and design decisions that determine the appearance of brickwork are considered under four main headings.

- **Bricks** their colour, texture, shape and size.
- **Mortar joints** their colour, texture, thickness and profile.
- **Bond pattern** resulting from the way the bricks in one course overlap or bond with those in the course below.
- **Detailed design** that enhances or enriches plain brickwork and architectural features.

But first a warning!

The appearance of a building is important simply because throughout its life it will be of major concern not only to its users but to innumerable passers by. But responsible clients, designers and builders will also wish to ensure that a building has adequate durability, rain resistance, strength and robustness. This book continually refers to these matters when commenting on appearance.

2.1 Types of bricks – clay, concrete and calcium silicate

That 91% of the current production of all bricks is made from some type of clay is a good enough reason for beginning with clay bricks. But clay bricks, being made from many types of clay, are also the most varied of the three types. Concrete and calcium silicate bricks, which account for 7% and 2% of production respectively, are made from more consistent materials. Because all three types have many characteristics in common it easier to describe clay bricks first and then highlight those characteristics that are special to concrete and calcium silicate bricks. Manufacture of the three types is outlined in Chapter 5.

2.1.1 Clay bricks – visual characteristics

The colour, texture and regularity of clay bricks, as well as other physical properties such as strength, durability and absorbency are determined mainly by the mineral composition of the clay or clays from which they are made. The composition of clays varies considerably between the many deposits throughout the country (**Plate 1**).

When bricks were first made choice was limited to soft clay deposits that were comparatively easy to 'win', process and transport with the simple resources available.

As demand increased and spread geographically, improved manufacturing techniques and transport facilities provided access to a greater variety of brick-making clays and hence a greater variety of bricks.

The manufacturing processes used for different clays greatly affect the physical characteristics of the bricks. See Chapter 5.

a) Classification

Three varieties of clay bricks are defined in British Standard BS 6100: Part 5, as Facing, Common or Engineering. The terms facing and common both relate to appearance only and not to other attributes such as strength or durability.

Facing bricks are specially made or selected to give an attractive appearance.

Common bricks are suitable for general construction work, with no special claim to give an attractive appearance. However, common bricks are sometimes specified to be laid with care to give a fairfaced finish (**Plate 11**).

Engineering bricks have a dense and strong semi-vitreous body, conforming to defined limits for water absorption and compressive strength. Engineering bricks should not be assumed to have the qualities of a 'facing' brick. But in practice they are often built fairfaced.

Proprietary names given for product identification usually contain an indication of the colour and texture, e.g. a smooth red or multi-rustic. They do not refer to other physical and usually measurable properties which are normally given in manufacturers' literature.

b) Colours

The colours of clay bricks range from pale and bright reds, through yellows, buffs and browns to blues, purples and greys (**Plate 12**).

The colour of a brick after firing is dramatically different from its colour in the 'green' state after it has been formed and dried ready for firing.

(i) Colour and the composition of clay

Brickmaking clays are composed mainly of alumina and silica, but other elements present in comparatively small proportions have a greater effect on colour. In general, the greater the proportion of iron oxide within the clay the redder the bricks will be. At about 7% the colour may be a deep purple. Finely ground chalk and limestone present with iron oxide, as in London clay, produces yellowish or buff London Stock bricks. Without the iron oxide the bricks would be whitish. Magnesium oxide tends to produce yellow bricks.

Gault clays produce a very pale grey or so-called white brick. Bricks from shales in coal-bearing strata, found mainly in the Midlands and the North, generally contain iron oxide and fire red. See Section 5.1.1.

(ii) Colour and the manufacturing processes

The 'natural' colour of the fired-clay body can be modified by admixtures introduced during manufacture. For instance manganese dioxide turns what would have been a red brick into a brown one.

The colour of the faces of moulded bricks may be determined by the type of sand used to separate the moulds from the sticky clay. The faces of extruded wire-cut bricks may be coloured by applying sand to the columns of clay before they are cut to shape. In both cases the colour of the sand can be changed by staining with pigments.

Bricks that have individual faces of two or more colours are called 'multi-coloured' or simply 'multis'. The colour variations they describe are generally the result of air supply and temperature variations within a kiln. The term 'brindles' is also used to describe similar bricks. As there are no precise definitions of these two terms, their application and interpretation may vary throughout the country.

The colour as well as the texture of bricks made by the soft-mud process may result from incompletely burned solid fuel that has been mixed with the clay to help it burn.

The control of air to restrict the flow of oxygen in a kiln provides a 'reducing' atmosphere which is used to produce blue bricks and some 'flashed' multi-coloured products.

c) Texture of clay bricks

The textures of clay bricks range from glazed and smooth, through lightly textured to rough and pitted surfaces (**Plate 12**).

(i) Moulded soft-mud bricks and 'hand-mades'

The earliest bricks were made from soft malleable or 'plastic' clays. Such bricks are still in demand for their unique appearance and possibly their association with many much admired older buildings. They have a fairly open texture derived partly from the relatively coarse sands used to aid separation from the moulds. Finer sands are used to produce smoother surfaces such as those required for gauged brickwork. See Section 2.4.4.

The texture of many of the soft-mud bricks both old and new is affected by the inclusion in some cases of partly burnt solid fuel to aid in the firing. This can result in quite rough or pitted areas (**Plate 9**).

The hand-making process provides a characteristic surface texture known as 'creasing'. This can be seen in **Plate 14**.

(ii) Smooth faces

When harder, deeper lying clays became economically accessible by modern 'winning' or excavating methods, the introduction of pressing by machinery, rather than hand-moulding, produced hard bricks with smoother faces that seem to

have been particularly popular with the Victorians. Typical of these bricks were those made from colliery shales plentiful in the Midlands and the North and particularly associated with industrial buildings.

These bricks, now generally made by high pressure extrusion through steel dies, are often used in modern buildings for their regular shape and sharp arrises to produce precise looking brickwork (e.g. **Plates 15, 16 and 18**).

(iii) Surface treatments

The smooth surface of an extruded column of clay can be modified by applying sand to it or cutting or scraping the smooth surface to roughen it, or dragging the coarser aggregate along the face.

One of the earliest surface treatments of modern bricks was the rustic finish applied to a pressed Fletton common brick to turn it into a facing brick. Nowadays, sands and texturing are applied for this purpose. See **Figure 2.1** which shows two 'common' bricks next to two facing versions of the same with textured and sanded faces. The latter rest on a red engineering brick.

(iv) Glazed bricks

Glazed bricks are made by coating the faces to be seen in the face work with a

Fig. 2.1 Facing, common and engineering bricks.

clay slip or ceramic glaze which is coloured and, when fired, produces a high gloss surface. For many people glazed bricks conjure up images of utilitarian white, dark brown or green glazed walls where ease of cleaning was paramount. But in the past and recent years architects have exploited the delight of coloured, glazed bricks both in large areas and in small quantities within a wall of contrasting bricks where they can have a jewel-like quality (**Plates 17, 19, 20 and 21**).

d) Sizes of clay bricks

Bricklaying is an efficient production technique which allows one hand to remain free to handle and position the bricks while the other continually applies mortar with a trowel. The elegant, flowing, rhythm of a skilled bricklayer enables the process to be maintained for long periods only because bricks are neither too wide nor too heavy.

For brickwork to compete as a versatile walling material the length to width to height ratios must facilitate not only regular bonding of normal brick walling but also the detailing of architectural features (**Fig. 2.2**).

We may assume that the present size has evolved as one that best suits the many requirements.

(i) Sizes through the ages

The historical study of brick sizes in Britain requires much patient research and is outside the scope of this book. But a few key facts are noted for interest.

- For the first 400 years AD the Romans in Britain made bricks of a wide variety of sizes and laid them in ways that had little in common with the brickwork with which we are now so familiar. See Section 1.1.
- 13th century bricks ranged from 8 to 12 inches long by 3¾ to 4¾ inches wide and 1¾ to 2 inches high.

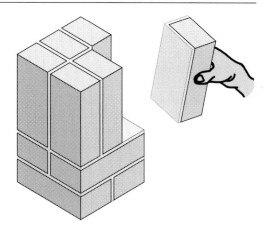

Fig. 2.2 Brick dimensions – provide versatility in design and efficiency in assembly.

- In the late 15th century, bricks 9½ × 4½ × 2 inches were common.
- In 1571 a brickmakers' charter established a size of 9 × 4½ × 2¼ inches which persisted until the end of the 17th century when a thickness of 2½ inches became general.
- An act of 1776 specified a minimum size of 8¼ × 4 × 2½ inches but this was to protect builders from unscrupulous brickmakers rather than to establish a standard size.
- In the late 18th century many bricks were around 3 inches high, possibly encouraged by the brick tax of 1784 which was levied per 1000 bricks irrespective of size. Increasing the brick size as a way of avoiding taxes was countered in 1803 by the government doubling the duty on bricks exceeding 150 cubic inches, which is the volume of a brick measuring 9½ × 4½ × 3½ inches.
- During the 19th century architectural fashion, especially in the North, Midlands, South Wales and Scotland, favoured bricks about 3 inches in height. Bricks 3⅛ inches high were used for the dockside buildings at Albert Dock in Liverpool, 1841–45.
- In 1904 the 'RIBA (Royal Institute of British Architects) Standard Sizes of Bricks' was published following dis-

cussions with the Brickmakers Association in consultation with the Institution of Civil Engineers. The maximum and minimum sizes were $9 \times 4\frac{3}{8} \times 2\frac{11}{16}$ inches and $8\frac{7}{8} \times 4\frac{5}{16} \times 2\frac{5}{8}$ inches allowing four courses to rise 12 inches. In the 1920s, following a conference which included the Northern Federation of Building and Engineering Brick Trades, a second RIBA Standard was published which had the same sizes as the first except that the maximum and minimum heights were $2\frac{15}{16}$ and $2\frac{7}{8}$ inches allowing four courses to rise 13 inches. In practice the two Standards became the 'Southern' and 'Northern' bricks respectively.

- The RIBA Standards were eventually incorporated in a British Standard. All this reflected a growing realization that although it would continue to be possible to find a supplier for special sizes of bricks there was no doubt of the convenience of a standard size for the majority of brickwork.

The foregoing demonstrates that the sizes of bricks in our national building stock vary considerably, but the casual observer may be more aware of the face proportions than the linear dimensions. For most people it is probably the 'thin' bricks, say 2 to $2\frac{1}{4}$ inches (50 to 57 mm) high and 'thick' bricks, say $2\frac{3}{4}$ inches (70 mm) high and more that are visually distinctive. For some, 'thin' bricks may conjure up images of medieval or Tudor times or of Dutch or Italian architecture.

'Thick' bricks may invoke the mills and other industrial buildings in the Midlands and the North of England.

It follows that dating buildings by brick size is not generally reliable although it may corroborate evidence of architectural style and other characteristics.

(ii) Modern bricks

There are three ways of describing the sizes of modern bricks. But be warned, the definitions have been revised in recent years as noted below (**Fig. 2.3**).

The **coordinating sizes** are those of the space allocated to a brick including allowances for joints and tolerances. The coordinating sizes of a brick are most easily thought of as being the measurement between the centre lines of nominal 10 mm wide joints either side of the brick. For standard bricks the coordinating sizes are $225 \times 112.5 \times 75$ mm. Note: The 'coordinating size' was formerly called the 'format size'.

The **work sizes** are the sizes specified for manufacture. They may be thought of as the 'target' sizes at which the manufacturer aims. In the case of bricks the work sizes are derived from the coordinating sizes less a nominal allowance of 10 mm for joint widths. For standard bricks they are therefore $215 \times 102.5 \times 65$ mm. Note: The 'work size' was formerly called the 'actual size'.

The **actual sizes** are now those of individual bricks as measured. They will invariably be a little more or less than the work sizes and will vary from brick to brick. See Section 2.1.1 f.

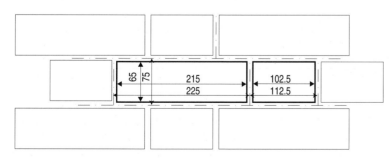

Fig. 2.3 Solid arrow heads indicate work sizes. Open arrow heads indicate coordinating sizes.

Standard size. Modern bricks are available in standard and non-standard sizes. The sizes of all bricks made to British Standards are listed in Table 2.1.

Currently in the United Kingdom, the great majority of clay bricks are made to the metric standard coordinating sizes of $225 \times 112.5 \times 75$ mm in BS 3921: 1985 which are about 2% less than the the the comparable Imperial standard sizes of $9 \times 4\frac{1}{2} \times 3$ inches in BS 3921: 1965.

In anticipation of the need for metric modular bricks, proposals for a number of different sizes were published in British Standard Drafts for Development. Eventually one standard modular brick with coordinating sizes of $200 \times$ 100×75 mm was recommended in BS 6649: 1985.

Within BS 4729: 1990 'Bricks of special shapes and sizes', Group NS consists of nine non-standard cuboid bricks. The first three in the group are alternatives to the modular standard brick of $200 \times 100 \times 75$ mm. The next four are of the metric standard length and width, two of which are 'thin' and two are 'thick'. The last two are longer, wider and higher than the metric standard brick and emulate the traditional larger 'Northern' brick.

Non-standard sizes. Despite the many sizes now defined in the various standards, a brickmaker may sometimes

Table 2.1

Type of brick	Length	Width	Height	British Standard
Imperial Standard	**9 in**	**4½ in**	**3 in**	BS 3921: 1965
	(228.6 mm)	(114.3 mm)	(76.2 mm)	
	8⅝ in	*4⅛ in*	*2⅝ in*	
	(219 mm)	*(104.8 mm)*	*(66.6 mm)*	
Metric Standard	**225 mm**	**112.5 mm**	**75 mm**	BS 3921: 1985 &
	215 mm	*102.5 mm*	*65 mm*	BS 187: 1978
Modular Standard	**200 mm**	**100 mm**	**75 mm**	BS 6649: 1985
	190 mm	*90 mm*	*65 mm*	
Standard 'Specials'				BS 4729: 1990
Group NS 1.1	*190 mm*	*90 mm*	*90 mm*	⎫
Group NS 1.2	*290 mm*	*90 mm*	*65 mm*	⎬ Alternative modular bricks
Group NS 1.3	*290 mm*	*90 mm*	*90 mm*	⎭
Group NS 1.4	*215 mm*	*102 mm*	*50 mm*	⎫ 'Standard' length & width
Group NS 1.5	*215 mm*	*102 mm*	*53 mm*	⎬ but 'thin'
Group NS 1.6	*215 mm*	*102 mm*	*73 mm*	'Standard length & width but
Group NS 1.7	*215 mm*	*102 mm*	*80 mm*	high to course with old 'Northern' bricks.
Group NS 1.8	*233 mm*	*112 mm*	*73 mm*	Large bricks to emulate
Group NS 1.9	*233 mm*	*112 mm*	*80 mm*	'Northern' traditional bricks.

Notes: 1) Coordinated sizes are in **bold** e.g. **9 in** and **225 mm**. 2) Work sizes are in *italics* e.g. *215 mm*. 3) Metric equivalents of imperial dimensions are in brackets e.g. (228.6 mm). 4) in BS 3921: 1965 & 1969 'Coordinating sizes' were referred to as 'Designation', and 'Work sizes' as 'Actual dimensions'. 5) Group NS (non standard) refers to cuboid bricks not included in the Metric Standard BS 3921 or the Modular Standard BS 6649. It is in that respect they are non-standard.

be asked for a brick of non-standard dimensions. At the new Glyndebourne Opera House (**Plates 22–24**) completed in 1994, approximately 1.5 million machine-made, hand-finished bricks were made to work sizes of 220 × 106 × 60 mm high (8⅝ × 4⅛ × 2⅜ inches) to match those in an existing house nearby.

For the Inland Revenue buildings in Nottingham, the same architect used a brick with work sizes of 230 × 112.5 × 75 mm (9 × 4⅜ × 3 inches) laid with nominal 5 mm joints to echo the character of the brickwork in nearby 19th century warehouses.

It is clear that, although the sizes of bricks have been standardized to enable costs to be reduced by rationalizing the greater part of brick production, the brickmaking industry has retained the flexibility to meet special requirements.

e) Regularity of shape of clay bricks

Although most modern clay bricks have the highly regular shape often thought appropriate for modern brickwork some inherently irregular bricks are still considered attractive and made and used where appropriate. Specific limits on bowing and twisting are not specified by British Standards but, in practice, limits are imposed by the permitted deviations in length, width and height.

The regularity of bricks is determined largely by the forming, drying and firing of the clay. For example, moulded bricks are inherently less regular at the 'green' stage than extruded and wire-cut and machine pressed bricks. Subsequently, irregularity is minimized by very careful control of the drying process as well as the rate of firing and the ultimate kiln temperature.

Temperatures can be controlled within fine limits in the most modern computer controlled tunnel kilns. By contrast clamps and simple intermittent kilns cannot be so precisely controlled, and so usually produce some overfired and distorted bricks.

Because bricklayers are generally trained to build to line, level and plumb using the more regular types of bricks they may initially have difficulty in laying irregular bricks until they have some site experience. However, it has been known for an architect to have difficulty in persuading bricklayers not to discard the more irregular bricks which he felt dramatically expressed the nature of the bricks and the way they were made.

f) Accuracy of clay bricks

Bricks differing by 2 inches in length have been noted in 15th century brickwork. Such gross deviations in size became unacceptable once brickwork developed as a facing material and was required to have a regular bond pattern.

Attempts to standardize limits to deviations in size culminated in a table of 'acceptable tolerances' for the length, width and height of standard bricks in BS 657, first published in 1950. The method of testing bricks for conformity with the Standard is still in use, and requires the overall measurement of 24 bricks rather than the measurement of individual bricks. This method may change with the adoption of European standards.

As most brickwork can be built satisfactorily with bricks that comply with the limits of size permitted in BS 3921, it would be unnecessarily expensive to specify bricks of greater accuracy. A good bricklayer will slightly vary the width of vertical cross-joints from the nominal 10 mm but will ensure that they are reasonably consistent. It is a matter of judgement in particular circumstances. Variations in the width of cross-joints are usually less acceptable between regular looking bricks than between stock bricks.

When building brickwork in which bricks complying with the standard limits of size would be unacceptable, for example in narrow piers with few cross-joints (**Fig. 2.4**), in soldier courses or in brick cappings across a one-brick thick wall, it

Fig. 2.4 A narrow pier with only two joints in a course.

will be necessary to select the bricks so that there is no more than say a 2 mm difference between the lengths of bricks. For large quantities this is best done before delivery by prior agreement with the supplier, otherwise the bricklaying team should be instructed to do this on site. See Sections 3.2.6, 2.4.2 a and 2.4.3 b.

g) Brick shapes

How can the shape of a brick be described other than 'brick shaped' which is tautologous? In this book the term 'rectangular cuboid' is used as the neatest way of describing a six-sided figure in which opposite sides are equal rectangles and is easier to remember than 'parallelepiped' which means the same.

Bricks have been made this shape for over 6000 years, presumably because it is the most practical and economic shape to manufacture, handle, transport and build. The shape has an enduring and satisfying appearance born of simplicity and fitness for purpose.

This does not deny the need for 'special' shapes in order to build refined, elegant and decorative details.

(i) Standard bricks of special shapes and sizes

The earliest brickmakers and builders must have soon realized that specially shaped bricks could be readily made to simplify and quicken the building of more robust and elegant brickwork than by cutting bricks to shape on site. Over the years many shapes have been developed to meet special needs and their use has for long been a major feature of brickwork.

Towards the end of the 1980s a review of the British Standard for special shaped bricks concluded that there was a need to amend and extend the range in view of the increased use of special shapes during the previous decade and because many of the dimensions had been determined to suit solid rather than cavity walls. The revised standard was published as BS 4729: 1990 'Dimensions of bricks of special shapes and sizes'. They are divided into ten major groups containing over 300 variations.

Because most 'specials' require special manufacturing techniques it is advisable for anyone proposing to use them to discuss their requirements with a manufacturer as soon as possible to ensure that their needs are met. If none of the standard 'specials' will suit the user's needs, manufacturers can, given sufficient notice, produce purpose-made non-standard special shapes.

(ii) Groups of 'specials' in BS 4729: 1990

The current standard Special shaped bricks are grouped in the current standard according to the way they are intended to articulate brickwork. They are dimensioned to complement the $225 \times 112.5 \times 75$ mm standard bricks.

Availability Many manufacturers stock quantities of the more commonly used standard special shapes to match their main facing products. Otherwise, specials are made to order. Some manufacturers, because of the nature of their particular clay or manufacturing processes, are not able to produce all the shapes, especially the very large ones.

Choice of specials The way most special shapes are intended to be used is obvious. But users who have any doubts about the choice of a special shape to meet their needs should consult the manufacturers as soon as possible. Many have computer aided design facilities to assist in advising designers.

Matching standard bricks Again, it is generally advisable for customers to discuss their total requirements at an early stage to enable the manufacturer to ensure that the specials match the standard bricks in colour and texture as they are often made by different methods.

Faced surfaces The surfaces of standard special bricks that are faced are identified in the British Standard or manufacturers' literature (**Fig. 2.5**).

Left- and right-handed variants The geometry of some asymmetric shapes like 'plinth external returns', obviously precludes their use on the opposite hand merely by turning them upside down. At locations such as either side of a window or door opening or at a

corner that has to be bonded, handed versions are essential (**Fig. 2.6a**).

The geometry of some simple asymmetric shapes like a single cant will allow it to be handed merely by inverting it, (**Fig. 2.6b**); however, this may be visually unacceptable with some types of bricks which have textured surfaces that appear quite different when laid 'upside down' (**Fig. 2.6c**). Also if bricks must be

(a)

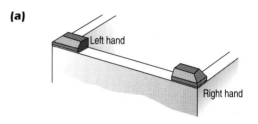

(b)

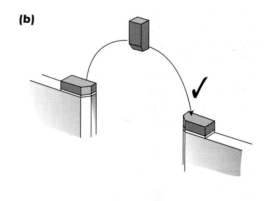

(c)

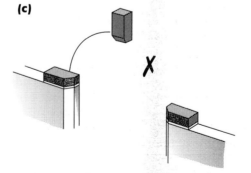

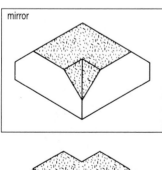

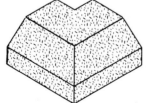

Fig. 2.5 Not all surfaces of 'specials' are faced.

Fig. 2.6 'Special bricks – left- and right-handed variants. (a) external plinth returns are not reversible, so must be handed (b) smooth cant bricks may be reversible (c) textured cant bricks may not be reversible.

laid frog up in order to support concentrated loads, such as the end of lintels, they cannot be handed by turning upside down unless the frog is first filled with mortar before being laid.

Tolerances   It has proved difficult to specify dimensional tolerances for individual 'special' bricks. In practice, consistency of linear dimensions is less important than consistency of shape such as the angles of 'angle' and 'cant' bricks, or the curvature of bullnose bricks. This is particularly important where they are to be handed and bonded at corners.

As these requirements are not covered by the British Standard specification the matter should be dealt with by early discussion with the manufacturers so that they are fully aware of any particular requirements for a project.

Further information   A detailed explanation of special shapes and their use is provided in a well illustrated design note, DN 13, 'The Use of Bricks of Special Shape' by the Brick Development Association.

(iii) Bricks of special shapes and sizes specified in BS 4729: 1990

Copings and cappings – group CP
Two standard 'coping' and two 'capping' units both in half-round and saddleback form are sized for use atop one-brick thick walls (**Fig. 2.7**). The addition of capping units in the revised standard reflected the popularity of flush detailing during the 1970s and 1980s.

The Standard includes no stop ends or returns for these units to enable the ends to be more stable and attractive, but generally manufacturers can supply them purpose-made. See Section 2.4.3 b.

Bullnose bricks – group BN   'Bullnose' bricks enable arrises vulnerable to damage to be avoided by providing robust rounded profiles. These may be either vertical at corners and reveals to doorways, or horizontal along the edges of cappings or sills. Visually, they impart

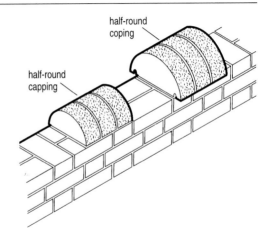

Fig. 2.7 Standard half-round capping and coping units.

a less angular but more robust appearance (**Fig. 2.8**).

Some shapes provide bullnoses to brick-on-end or brick-on-edge cappings or sills. The bullnose may be on one side of a brick, a 'single bullnose', or on two opposite sides, a 'double bullnose'. 'Stops' provide a transition from one bullnose brick to a cuboid brick. Stop ends and returns provide robust and elegant solutions at ends and junctions of freestanding walls. This very extensive range provides the means of enhancing many features in brickwork (**Fig. 2.9**).

The Standard includes alternative bullnose radii of 51 mm, i.e. half the width of a brick, or 25 mm. Careful consideration should be given to the choice as the appearance, depending on the context, can be remarkably different.

Angle bricks – group AN   'Cant' bricks which were formerly called 'splay' bricks, provide similar features to the bullnose range but with a 45° cant or splay, instead of the bullnose. Similarly, there are 'single' and 'double cants' which are also complemented by a range of stop, stop end and external and internal return units (**Fig. 2.10**). There are two sizes of splay which correspond to and are derived from the splays on plinth (PL) bricks – see below.

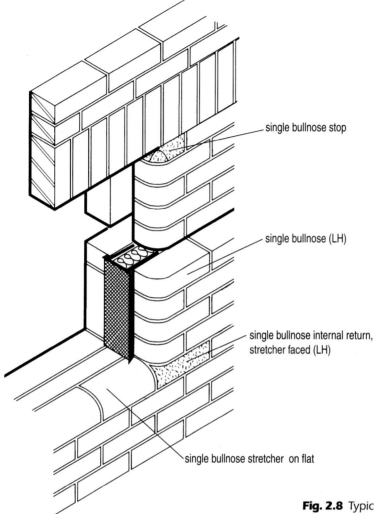

single bullnose stop

single bullnose (LH)

single bullnose internal return, stretcher faced (LH)

single bullnose stretcher on flat

Fig. 2.8 Typical use of bullnose bricks to form window opening.

The 'squint', 'external' and 'internal angles' in the AN group facilitate angles in brickwork of 30, 45 and 60 degrees in plan. For each of the three external and internal angles there are three possible solutions depending on the bond pattern required at the angle (**Fig. 2.11**). Illustrations in BS 4729 and BDA's Design Note 13 help users choose the most appropriate one for their application.

Plinth bricks – group PL   Plinth bricks were traditionally used to reduce a wall thickness above the base, thereby forming a weathered and attractive top to a plinth (**Fig. 2.12**). But in recent years they have often been used inverted in half-brick thick cladding to simulate corbelling. See Section 2.4.3 d. The vertical face can be either 23 or 9 mm high. See Section 2.4.3 e.

As the plinth brick profiles match those of cant bricks they can be used together, e.g. cant bricks at the reveals together with plinth bricks at the sills and heads. A special plinth stop or cant stop

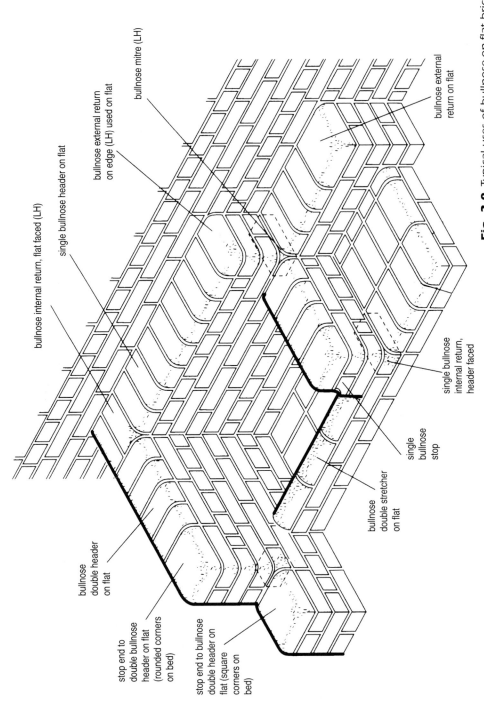

bullnose internal return, flat faced (LH)

single bullnose header on flat

bullnose external return on edge (LH) used on flat

bullnose mitre (LH)

bullnose external return on flat

single bullnose internal return, header faced

single bullnose stop

bullnose double stretcher on flat

bullnose double header on flat

stop end to double bullnose header on flat (rounded corners on bed)

stop end to bullnose double header on flat (square corners on bed)

Fig. 2.9 Typical uses of bullnose-on-flat bricks.

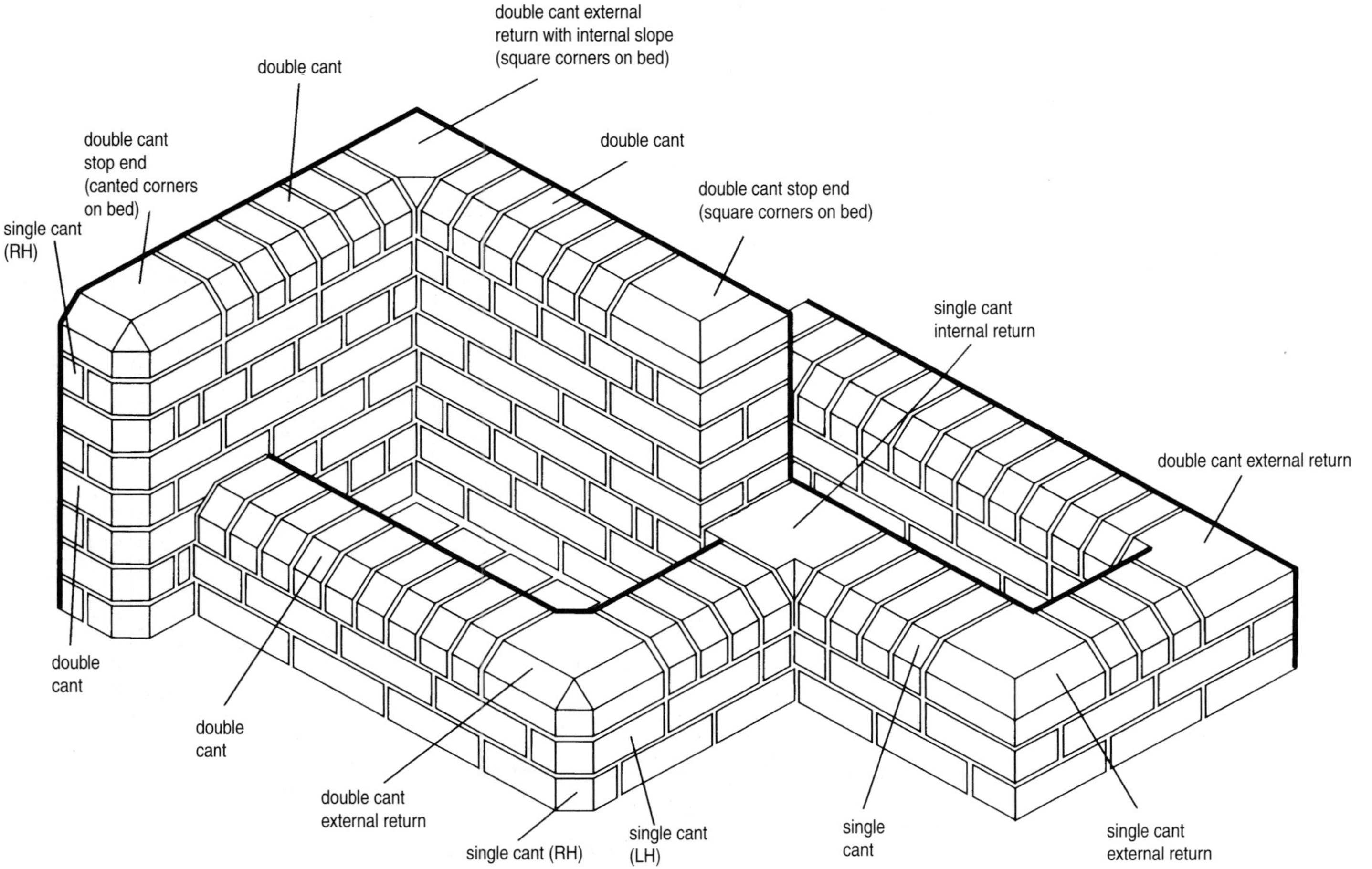

Fig. 2.10 Typical uses of cant bricks.

double cant external return

single cant
external return

single cant
internal return

single
cant

double cant stop end
(square corners on bed)

double cant external
return with internal slope
(square corners on bed)

double cant

double cant

single cant
(LH)

single cant (RH)

double cant
external return

double
cant

double cant
stop end
(canted corners
on bed)

single cant
(RH)

double
cant

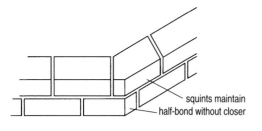

squints maintain
half-bond without closer

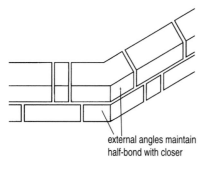

external angles maintain
half-bond with closer

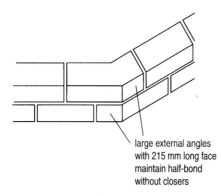

large external angles
with 215 mm long face
maintain half-bond
without closers

Fig. 2.11 Three alternative 'specials' for use at external angles.

provides the transition between the two (**Fig. 2.13**).

Arch bricks – group AR   The current Standard contains 'tapered headers' and 'tapered stretchers' as voussoirs to provide parallel sided joints between voussoirs for fairfaced circular arches spanning 4, 6, 8 or 12 standard bricks (900, 1360, 1810 and 2710 mm)(**Fig. 2.14**).

Similar bricks were referred to formerly as 'culvert radial side arch' and 'culvert radial end arch' bricks respectively – names which reflected their original civil engineering use in sewage culverts. See Section 2.4.3 c.

Radial bricks – group RD (Fig. 2.15) Curved brickwork makes a bold architectural form and is inherently structurally strong. The six 'radial headers' in the current standard are for building one-brick thick, fairfaced, convex walls in Header Bond and the six 'radial stretchers' are for half-brick thick, fairfaced, convex brickwork in Stretcher Bond. These radial bricks are dimensioned to suit the 'ideal' radii of 450, 675, 900, 1350, 2250 and 5400 mm.

Standard radial bricks allow neat construction of quadrants of the 'ideal' radii without cutting bricks for bonding with straight work.

The 1971 version of the Standard contained similar bricks with the additional names of chimney or well headers and stretchers.

For other aspects of designing and building curved walls see Section 2.4.3 a.

Brick slips – group SL   'Face slips' are thin slices of brick having standard face dimensions of 215×65 mm and 25, 30, 40, or 50 mm thick. In the 1960s and 1970s they were often bedded to hide the vertical faces of concrete edge beams supporting the facing brickwork.

Following failures of adhesion, caused largely by bedding expanding clay slips on to shrinking concrete frames, the method has been largely superseded by supporting the brickwork on stainless steel angles. 'Face slips' are used successfully in less onerous conditions particularly to face internal or sheltered walls.

'A bed slip' has the same dimensions as the bed face of a standard brick (215×102.5 mm) and is 25 mm thick. One side is faced.

Soldier bricks – group SD   This group facilitates the turning of internal

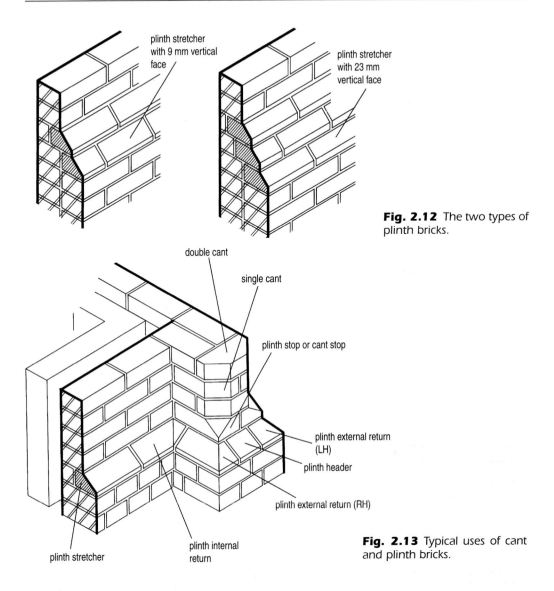

plinth stretcher with 9 mm vertical face

plinth stretcher with 23 mm vertical face

Fig. 2.12 The two types of plinth bricks.

double cant

single cant

plinth stop or cant stop

plinth external return (LH)

plinth header

plinth external return (RH)

plinth internal return

plinth stretcher

Fig. 2.13 Typical uses of cant and plinth bricks.

and external right angled corners in soldier courses of standard bricks and also of cant bricks used as 'soldiers'. See Section 2.4.2 a.

Non-standard cuboid bricks – group NS These are not special shapes but rectangular cuboid bricks having different dimensions from the 225 × 112.5 × 75 mm standard brick normally used. See Section 2.1.1 d (ii).

Some are larger than the standard

brick, and are for use where brick sizes, akin to the larger bricks traditionally used in the Midlands and North of England, are required. Some are 53 or 50 mm high to meet the occasional demand for 'thin' bricks.

Bonding bricks – group BD These bricks facilitate bonding without having to cut bricks on site. To the extent that they avoid inaccurate, rough-cut edges showing on the face, their use

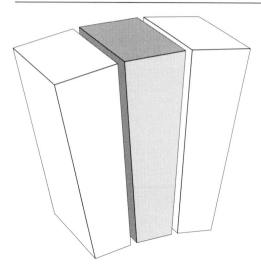

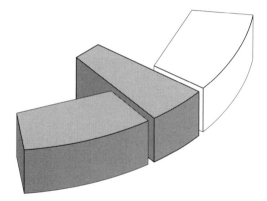

Fig. 2.15 Radial stretchers and header.

Fig. 2.14 Tapered stretchers. Tapered headers are also available.

will improve the appearance of brickwork where bonding or breaking bond requires a brick face other than a standard stretcher or header. See Section 2.3.1.

(iv) 'Cut and stuck' specials

As a matter of expediency, some special shapes have been produced by cutting standard bricks to appropriate shapes and joining them with epoxy resin adhesive. They are commonly known as 'cut and stuck' specials.

The technique is not uncommon and has produced successful results but requires skill and care in cutting accurately and preparing the surfaces to be stuck together. Some manufacturers do not approve of the technique at all, some do so for some of their products but with reservations. No manufacturer approves of the process being carried out on the construction site.

The appearance of 'cut and stuck' is generally detectable, especially after weathering and this may not be acceptable.

2.1.2 Visual characteristics of calcium silicate bricks

Approximately 2% of bricks made in the UK are calcium silicate (**Fig. 2.16**). Their composition and manufacture are described in Section 5.2.

a) Classification

Calcium silicate bricks are specified in accordance with BS 187: 1978 as facing bricks or loadbearing bricks.

Loadbearing bricks are required to be free from visible cracks and noticeable balls of clay, loam and lime and the compressive strength class may be specified.

Facing bricks are also required to be free from visible cracks and noticeable balls of clay, loam and lime but in addition shall be reasonably free from damaged arrises.

b) Colour

The natural colour of calcium silicate bricks is white but the addition to the mixture of highly colourfast iron oxide pigments creates a wide range of colours. The lime, being white, produces a particularly fine range of pastel colours. There is also a range of dark full-bodied colours and multi-colours.

Figure 2.16 Calcium silicate brickwork with a mural in contrasting bricks.

c) Texture

Normally the faces are exceptionally smooth as the result of the use of very fine aggregates which are partially dissolved to combine with the lime binder which is even finer than cement. A textured face can be produced by various surface treatments before or after curing.

d) Sizes and special shapes

All the brick dimensions in the current British Standards listed in Table 2.1 apply to calcium silicate bricks. The dimensions for calcium silicate bricks are specified in BS 187: 1978. The special shapes described in Section 2.1.1 g are applicable to calcium silicate bricks.

e) Regularity and accuracy

Accuracy and regularity are characteristic of calcium silicate bricks as no shrinkage takes place between pressing and curing the bricks. Generally, they are more consistent than clay bricks and concrete bricks, that is to say the actual sizes conform more closely to the work sizes. They have a precise appearance with sharp, crisp arrises.

f) Other characteristics

Calcium silicate bricks are not as prone to efflorescence as other types as they have a negligible soluble salt content. They are frost resistant and are available in five strength classes (numbered 3 to 7) ranging from 20.5 to 48.5 N/mm^2.

2.1.3 Visual characteristics of concrete bricks

Approximately 7% of bricks made in the UK are concrete. The composition and manufacture of these bricks are described in Section 5.3. Concrete bricks may be specified in accordance with BS 6073: Part 1: 1981 as facing bricks or loadbearing bricks.

a) Colour

The natural colour of concrete bricks is grey because of the predominance of ordinary Portland Cement. By the addition of pigments, which are claimed to be colour fast, concrete bricks can be made in a wide range of colours either mono- or multi-coloured.

b) Texture

Normally the faces have an even texture which may be modified by surface treatments before or after curing.

c) Sizes and special shapes

All the brick dimensions in the current British Standards listed in Table 2.1 apply to concrete bricks. The dimensions for concrete bricks are specified in BS 6073: Part 1: 1981. The special shapes described under clay bricks are equally applicable to concrete bricks.

d) Regularity and accuracy

Accuracy and regularity are characteristic of concrete bricks as little shrinkage takes place during manufacture. Generally, they are more consistent than most clay bricks, that is to say the actual sizes conform more closely to the work sizes.

They normally have a precise appearance with sharp, crisp arrises.

e) Other characteristics

Concrete bricks are not as prone to efflorescence as some types of bricks because they have a negligible soluble salt content. However, if newly manufactured units are subjected to very wet conditions, lime bleeding can occur and this is sometimes referred to as efflorescence. They are frost resistant.

2.2 Mortar joints – how their design and specification affects the appearance of brickwork

The appearance of brickwork, viewed from near or far, is dramatically and often surprisingly affected by the colour, thickness, profile and texture of mortar joints. But this should not be so surprising as joints generally account for 20% of facing brickwork.

The appearance of proposed new brickwork can be difficult to anticipate by looking at a single brick or even a small sample panel of slip bricks. It is better to inspect existing brickwork built from the same brick, carefully noting the colour, texture and profile of the mortar joints.

If possible, a trial panel should be built, but time must be allowed before inspection as the colours of the bricks and particularly the mortar will lighten as they dry.

The various aspects of mortar joints that affect the appearance of brickwork are referred to below.

2.2.1 Joint colour

The perceived colour of brickwork changes dramatically with the colour of the mortar joints as can be seen in the wall under construction at Bushey Park near to Hampton Court, where part only had been pointed (**Fig. 2.17**).

Consistently coloured mortar joints depend on good site management, supervision and care and attention by the bricklaying team. See Section 3.1 4.

The dramatic effect of mortar colour can be used to form patterned brickwork. Before the jointing mortar sets hard it is raked to a depth of about 15 mm and later pointed with coloured mortars (**Plate 25**).

Unintentional changes of mortar colour as the result of poor site control can be quite unattractive and may be unacceptable.

2.2.2 Joint thickness

In practice the thickness and regularity of mortar joints has an equally profound effect on the appearance of brickwork (**Figs. 2.18** and **2.19**).

Mortar joints in the earliest brickwork were considerably thicker than those to which we are accustomed today because many of the bricks were uneven in shape. The result was appropriate for the informal and picturesque architecture of the time. During the first half of the 17th century bricks became less variable but

Fig. 2.17 *A wall under construction – only partly pointed.*

Fig. 2.18 Thick mortar joints with 'thin' bricks.

Fig. 2.19 Fine mortar joints.

brickwork still had relatively thick joints.

With the production of more regular and accurate bricks, jointing techniques were developed to produce thinner joints and a more regular and refined brickwork better suited to the more formal and refined architecture of the time. The brickwork in the addition to Hampton Court Palace by Christopher Wren and

that in the adjacent original Tudor palace provide an interesting comparison (**Plate 26**).

Today, brickwork using standard bricks which have a worksize of 65 mm are normally gauged vertically at four courses to 300 mm having bed joints nominally 10 mm thick. The actual thickness will depend on the actual height of the bricks. See Section 2.1.1 d(ii).

2.2.3 Joint profile and texture

When a brick is bedded, the surplus mortar that is squeezed out is immediately cut off flush leaving an open textured surface to be finished when partly set. The way joints are finished affects the appearance of brickwork almost as much as their colour.

'Jointing-up', as the process of finishing a joint is known, should be co-ordinated so that identical tools and techniques are used to maintain uniformity of appearance.

With few exceptions most of the finishes may be produced by either 'jointing' or 'pointing'. The four most common joint profiles are shown in **Fig. 2.20**.

a) Flush joints

These are formed by cutting off excess mortar with the trowel and leaving the surface to set and harden, but they are normally finished either with the end grain of a piece of wood or coarse cloth. The latter is sometimes referred to as a 'bagged' joint and is often considered an appropriate joint finish to use with handmade or stock bricks.

b) Bucket handle joints

Bucket handle joints have a concave, slightly recessed profile formed by drawing a rounded section tool along the joint. Originally an old metal bucket handle was used but now it is more likely to be a piece of hosepipe or a purpose-made tool. It is commonly used today,

Fig. 2.20 Common joint profiles – top downwards – 'flush'; 'bucket handle'; 'recessed'; 'struck weathered'.

presumably because it is easy to form and, although rather bland, generally acceptable.

c) Recessed joints

These provide a dramatic shadow effect which, from close to, tends to emphasize the individual bricks rather than the brickwork as a whole. However, they do saturate the edges of the bricks with the increased risk of rain penetration and frost attack. The masonry code of practice, supported by most manufacturers, does not recommend the use of recessed joints in exposed conditions and particularly with bricks that are moderately frost resistant.

d) Weathered joints

Also known as weather struck joints these were introduced in the latter part of the 19th century and are amongst

the most common types of joint used today.

The vertical joints are normally struck before the bed joints with the left-hand side of the joints pressed back (See **Fig. 3.12**). This should be done consistently by both left- and right-handed bricklayers otherwise the light will be reflected differently from the cross-joints formed by each, causing the brickwork to appear patchy.

The bed joints are finished with the mortar flush with the top of the brick course beneath and typically about 2 mm back from the bottom of the bricks above creating a slight shadow. Anything more may make the finish look coarse. The metal trowel imparts a smooth, firmed finish to the mortar and improves the rain resistance of the joint.

Four less common joints are shown in **Fig. 2.21**.

e) Overhand struck joints

Also known as 'reverse struck', these joints were apparently developed at the end of the 17th century. It has been said that they are so called because they are indicative of the type of joint which would be naturally formed by a bricklayer leaning over a wall from behind and thus building overhand. However, it seems that use of the joint was not confined to overhand work. The appearance differs from the weathered struck joint because there is no shadow line under the bricks.

Nowadays its use is not universally recommended as the small ledge formed along the tops of the bricks becomes saturated which in exposed areas increases the risk of rain penetration and frost attack on moderately frost resistant bricks.

f) Double struck joints

These are said to have been commonly used from Tudor times until the 17th century. Presumably, few joints would have survived the years of weathering, and

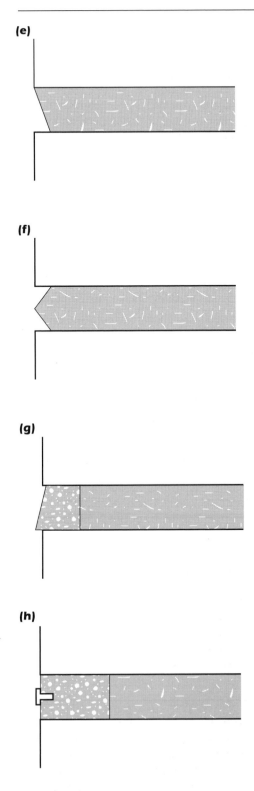

Fig. 2.21 Other joint profiles – top downwards – 'overhand struck', 'double struck', 'weather struck and cut pointing', 'tuck pointing'.

most would have been repointed using a less labour intensive method. A well preserved example is to be seen in the great kitchens of Hampton Court Palace.

g) Weather struck and cut pointing

A commonly used profile for pointing and repointing. It is not practicable to form the profile as part of the jointing process. The process of pointing is described and illustrated in detail in *The BDA Guide to Successful Brickwork*. Section 2.8 'Pointing and repointing'.

h) Tuck pointing

Tuck pointing came into use in the early 18th century and was commonly used until the end of the 19th century. It was a means of transforming the commonly used brickwork of irregular stock bricks and wide mortar joints to have something of the appearance of gauged brickwork in which joints are only a few millimetres thick. See Section 2.4.4.

The jointing mortar was raked out and pointed with a mortar which matched the colour of the bricks. Grooves were made in the soft mortar to form an accurate and regular joint pattern. To achieve this where the brickwork was very irregular the grooves might be cut into bricks where necessary and might not always follow the true joint.

The grooves were then usually filled with a fine line of lime putty which was usually white but sometimes red or yellow depending on the type of brick.

This type of work is highly specialized and today is mostly confined to the restoration of historic buildings.

2.2.4 Jointing and pointing brickwork

'**Jointing**' describes the process whereby bricklayers finish the face of the mortar joints as soon as the mortar is firm enough but before it hardens. Most brickwork today is built by jointing because it is cheaper even though any pigment required has to be added to the whole mortar mix. It is generally considered to be more robust than pointing. See Section 3.2.4.

'**Pointing**' describes the process whereby bricklayers rake out the joints to a depth of 15 to 18 mm before the mortar has hardened, and then fill or 'point' them at a later date with a mortar specially prepared for its appearance.

It is thought that the practice of pointing began sometime during the 18th century with the desire for more refined looking brickwork. It was considered the best practice to rake out the lime mortar and point the joints with a cement mortar which was considered superior though more expensive. This would not be considered good practice today.

Pointing requires knowledge, skill, experience and patience and is best done by a specialist. The appearance of attractive old brickwork can be spoiled by insensitive repointing and old bricks have deteriorated badly as the result of inadequate or inappropriate repointing.

For new work, pointing is labour intensive, costs more than jointing and is generally reserved for special effects where different colours are used in one wall or to minimize the risk of colour variation.

Repointing is often done to replace original pointing or jointing that has eroded to such a depth that it is unsightly. Mortar in poor condition may increase the risk of rain penetration or, less likely, it may lead to instability.

External and internal angles As external angles both in jointing and pointing are particularly noticeable they need to be neatly formed. Internal corners should emphasize the tie bricks or bonding at these points, finishing alternately to left and right not with a straight joint (**Fig. 2.22**).

2.3 Brickwork

Section 2.1 is about the colours, textures and forms of bricks, and Section 2.2 is about the colours and profiles of mortar joints. This section is about the appearance of brickwork when bricks and mortar are brought together.

2.3.1 Bonding brickwork

Bonding is the arrangement of bricks whereby those in one course overlap those in the course below, both along the length of a wall and across its width. The purpose of bonding is to make the

Fig. 2.22 Finishing internal angles: (left) a correctly finished internal angle; (right) an internal angle incorrectly finished with a straight joint.

brickwork homogeneous and to maximize its strength and robustness. It follows that the vertical cross-joints visible on the face of the brickwork in one course do not coincide with those in the course immediately below (**Fig. 2.23**).

The masonry code of practice recommends that 'the horizontal distance between cross-joints in successive courses (of bricks) should normally be not less than one quarter of the length of the bricks but in no case less than 50 mm'. In practice most brickwork is either quarter-bonded, half-bonded or both. More rarely it may be third-bonded.

Most early brickwork, being structural, was bonded, but because bricks varied considerably in length the bond pattern generated was irregular. This did not matter as brickwork was usually rendered, plastered, or faced with stone or marble.

As manufacturing techniques improved and bricks became more regular in shape it must have become evident that brickwork had the potential to be an attractive and economical facing material. The development of regular bond patterns was surely a refinement to improve the appearance of brickwork and provide an alternative to other masonry materials.

Until the end of the 1930s most brick walls were a minimum of one-brick thick and in either English or Flemish Bonds. Variations on these have developed in order to reduce labour and costs or to allow for steel reinforcement while the appeal of others may have been their decorative pattern. Some of these are illustrated in Section 2.3.2 below.

The long faces of bricks showing on the face of a wall are known as 'stretchers' and the short faces as 'headers'. The latter are normally the ends of 'tie' bricks laid across a wall.

Some specific craft aspects of bonding brickwork are described in Section 3.2.

a) Spreading loads

Bonding spreads vertical loads from walls, floors, roofs, beams and lintels over larger areas of brickwork (**Fig. 2.24**). In practice padstones of cast stone or concrete are often placed under concentrated loads such as the end of a beam to avoid localized crushing of the bricks.

Unbonded brickwork cannot spread these loads because the tendency of the loaded section to slide down between those on either side may not be adequately resisted by the continuous vertical mortar joints (**Fig. 2.25**).

Bonding also helps to resist wind and other lateral loads. When a lateral load is applied to a wall it will tend to deflect or bend. The bonding will help the wall to resist the tendency to fail about the vertical axis in what is called the 'strong' direction by spreading the loads horizontally to vertical strong points such as piers and columns along their length (**Fig. 2.26**). But, in the 'weak' direction, there is no bonding to help resist the tendency to bend about a horizontal or bed joint.

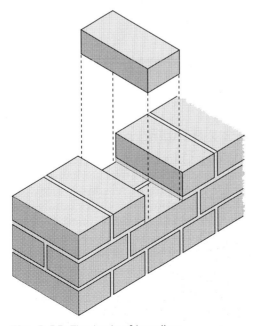

Fig. 2.23 The basis of bonding.

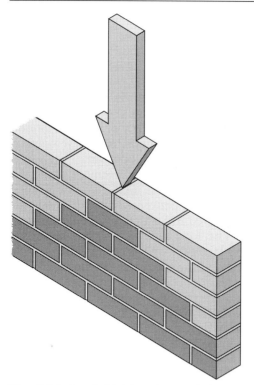

Fig. 2.24 Bonded brickwork spreads loads.

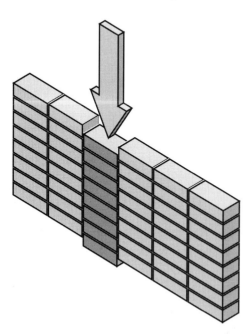

Fig. 2.25 Brickwork not bonded cannot spread loads.

b) Continuous vertical joints and brickwork strength

Continuous vertical joints can weaken brickwork especially if they extend to the outer face, for instance at a junction (**Fig. 2.27**). However, the limited vertical joints within a wall which are characteristic of some types of bond do not in practice noticeably weaken the brickwork.

For instance, it has for long been postulated that because English Bond (**Fig. 2.28a**) has no continuous vertical joints it is stronger than Flemish Bond (**Fig. 2.28b**)which has some continuous vertical joints. But experience and a limited number of relevant tests suggest that in practice there is no discernible difference. The structural masonry codes of practice, to which most engineers refer when designing structural brickwork, do not differentiate between types of bond when determining the permitted stress of masonry.

c) Rules of bonding

Today, the bonding of brickwork is as important from the point of view of appearance as it is for robustness and so bricklayer trainees are taught specific rules and principles that are fundamental to the craft of bricklaying.

Appropriately, the 'rules' are set out in Section 3.2.1, but an awareness of them will add to everyone's understanding and appreciation of brickwork.

2.3.2 Bonding patterns

Although most modern brickwork is a half-brick thick in Stretcher half-bond, it is, for a number of reasons, worthwhile noting the various brickwork bonds used over the years for walls one-brick and more thick. In hard landscaping, free-standing and earth-retaining walls are frequently built in solid brickwork. Furthermore, for some years, architects and engineers have shown an increasing interest in using structural brickwork both

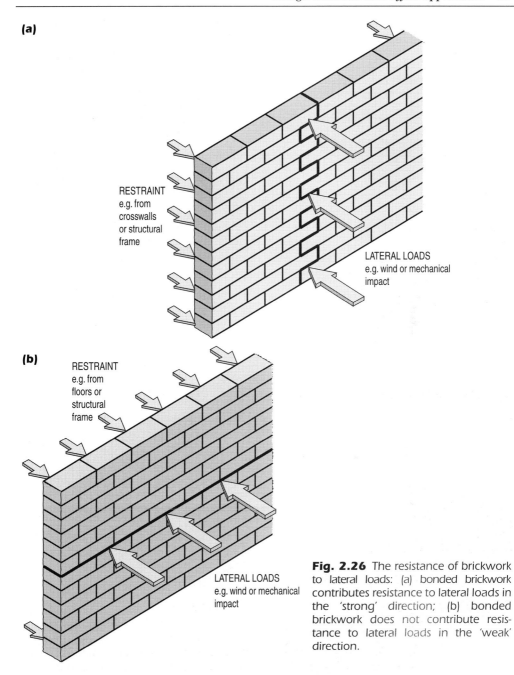

(a)

RESTRAINT
e.g. from
crosswalls
or structural
frame

LATERAL LOADS
e.g. wind or mechanical
impact

(b)

RESTRAINT
e.g. from
floors or
structural
frame

LATERAL LOADS
e.g. wind or mechanical
impact

Fig. 2.26 The resistance of brickwork to lateral loads: (a) bonded brickwork contributes resistance to lateral loads in the 'strong' direction; (b) bonded brickwork does not contribute resistance to lateral loads in the 'weak' direction.

internally and externally. There is also a growing desire amongst specialists and the general public that the best of our old buildings should be conserved with a sensitivity based on knowledge and understanding of the way they were built.

Several bond patterns are illustrated and described below. In each case the basic bond pattern is described first, followed by a description of the bricks at quoins, stop ends and reveals that generate the pattern.

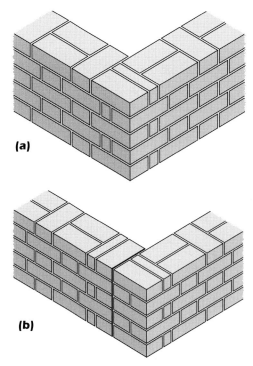

(a)

(b)

Fig. 2.27 Bonding at returns and junctions contributes to structural stability: (a) bonded return wall; (b) unbonded return wall showing straight joint.

a) English Bond

English Bond consists of alternate courses of stretchers and headers. The quoin header followed by a queen closer begins each header course. Generally, each stretcher course begins with a stretcher (**Fig. 2.29**).

The persistent rhythm of the header courses contrasting with the stretcher courses gives a horizontal emphasis.

The revival of brickwork in England in the 12th and 13th centuries began in the eastern and south eastern counties and was much influenced by brickwork in northern Europe including the Low Countries and Germany with whom a flourishing trade was developing. The bonding of much of the earliest brickwork was somewhat irregular, as at Beverley Bar, 1409–10, and had a large proportion of stretchers (**Fig. 2.51**).

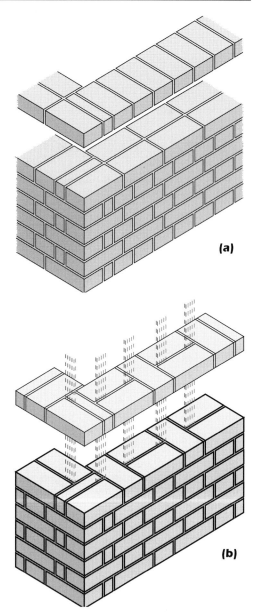

(a)

(b)

Fig. 2.28 A comparison between English and Flemish Bonds: (a) English Bond – showing lack of inherent continuous vertical joints; (b) Flemish Bond – showing extent of inherent continuous vertical joints.

Regular English Bond began to be used in England as at Tattershall Castle, 1431–49, and continued to be commonly used until the introduction of Flemish Bond about the second quarter

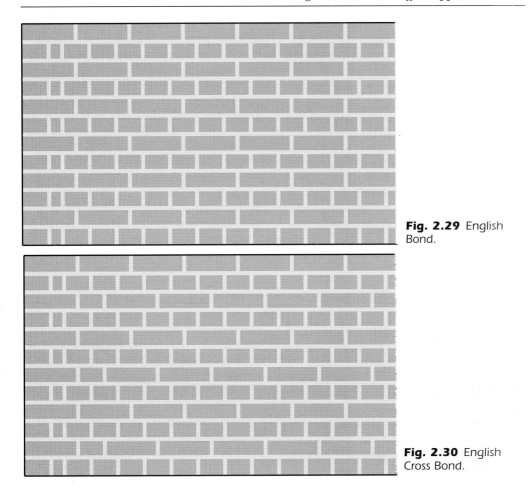

Fig. 2.29 English Bond.

Fig. 2.30 English Cross Bond.

of the 17th century. It has been suggested that it was at this time that English Bond began to be so called to distinguish it from the then new and 'foreign' Flemish Bond.

English Bond has for long been considered stronger than Flemish Bond because, unlike Flemish Bond, it has no continuous vertical joints from one course to the next and was generally used for engineering structures and large buildings such as 19th century railway works, canals, factories, warehouses and prisons. See Section 2.3.1 b.

English Bond was the norm for Dutch buildings of all periods although the Dutch call it 'Staand Verband' or 'Upright Bond'.

b) Variations

(i) English Cross Bond

This bond is like English Bond except that stretchers are positioned over the cross joints between stretchers two courses below. In other words alternate stretcher courses break joints (**Fig. 2.30**).

As in English Bond the quoin header followed by a queen closer begins each header course. In alternate stretcher courses a header is laid next to the first stretcher. As a result all courses rake back diagonally by a quarter of a brick giving some 'movement' to the pattern.

The bond which makes possible the introduction of small crosses and diaper pattern was and still is popular on the Continent. It has not been used extensively in England but is most often found in 16th century buildings. Layer Marney in Essex (about 1523) is one example.

(ii) Dutch Bond

Used in Germany and Holland in the 15th century Dutch Bond is identical to English Cross Bond except that each stretcher course begins with a three-quarter bat thereby avoiding the use of queen closers next to the quoin headers. A header is introduced into alternate stretcher courses. The illustration shows a repeating pattern picked out in dark bricks (**Fig. 2.31**).

(iii) English Garden-Wall Bond (Fig. 2.32)

This generally consists of one course of headers to three, or less often, five courses of stretchers. Each header course is quarter-bonded with the stretcher course below while the stretcher courses are half-bonded with each other, uniquely giving some half-bond and some quarter-bond in the same wall. The courses are generated in the same way as those in English Bond.

The bond has a number of variations and alternative names such as Scotch Bond, 3 and 1 Bond and, in America,

American or Liverpool Bond. It became popular in the 19th century and is commonly found in the United Kingdom.

English Garden-Wall Bond has an advantage over English Bond when building one-brick thick walls if both sides are to be fairfaced, as is often the case with boundary and other free-standing walls, hence its name. To achieve two fair faces the header bricks must be selected so that only those that do not vary greatly in length are used as through-the-wall headers. Obviously with fewer headers in a wall there will be less sorting and waste.

It can be argued that walls built in English Garden-Wall Bond are weaker than those built in English Bond because of the reduced number of headers or tie bricks and the continuous vertical joints between the stretchers in three or more successive courses. Although this is not known to have been confirmed by experience or tests, it would be prudent to consult a structural engineer before using the bond to build a wall that is to be heavily loaded. In this respect it is worthy of note that the Welsh Beck Granary, a large Bristol warehouse about the height of a six or seven storey block of flats, was built in English Garden-Wall Bond in 1869. It survives (**Plate 27**).

The bond was used in the external

Fig. 2.31 Dutch Bond – with a repeating pattern picked out in dark bricks.

Fig. 2.32 English Garden-Wall Bond.

walls of churches by Sir Giles Gilbert Scott in Northfleet, Kent; Luton, Bedfordshire; and Golders Green, London.

Stone-faced houses built about 1870, in Bristol, had side walls of brick in 3:1 English Garden-Wall Bond while other houses had a 5:1 version.

References have been found stating that English Garden-Wall Bond had been found suitable for large, brick furnace, chimney shafts. It is thought that the extra courses of half-bonded stretchers provided improved resistance to tensile forces around the circumference.

(iv) Facing or Common Bond (Fig. 2.33)

Like English Garden-Wall Bond in that generally it consists of one course of headers to three courses of stretchers. Also, the header courses are quarter-bonded with the stretcher courses below but there the similarity ends as the stretcher courses are quarter-bonded rather than half-bonded with each other.

The difference is generated by queen closers laid after the quoin headers in the second of the three stretcher courses as well as in the header courses. The result is a slightly more restless pattern than English Garden-Wall Bond.

(v) Raking English Garden-Wall Bond (Fig. 2.34)

This is also like English Garden-Wall Bond but in addition to the extra queen closer in the Facing Bond there is a header after the stretcher in the topmost of the three stretcher courses. The result is that the stretcher courses rake back diagonally by a quarter of a brick at a time giving much more movement to the pattern.

c) Flemish Bond (Fig. 2.35)

Flemish Bond consists of alternate headers and stretchers in every course. The headers in each course are laid centrally over the stretchers in the course below.

The pattern is generated in the same way as English Bond. The quoin headers followed by a queen closer begin alternate courses while the courses between begin with a stretcher. It produces a calm overall pattern with no strong direction.

It is probable that the majority of brick buildings built in Britain during the last three centuries have been in Flemish Bond. It has been suggested that it was adopted during the latter part of the 17th century because it was more suited than English Bond to the increasingly popular classical architectural style. But this interpretation is still open to question.

Fig. 2.33 Facing Bond.

Fig. 2.34 Raking English Garden-Wall Bond.

Fig. 2.35 Flemish Bond.

Although Flemish Bond was probably introduced from Flanders it was seldom used there except to achieve a chequer pattern by the use of headers of a different colour (**Fig. 2.36**). In Germany, Flemish Bond is known as Polinischer Verband or Polish Bond.

It is generally accepted that Kew Palace (**Plate 6**), built in 1631 by a city merchant of Dutch extraction, is one of the earliest examples of Flemish Bond in England.

From that time it increased in popularity in England until, in the 18th century, it was the most commonly used bond. A visit to Hampton Court Palace provides an opportunity to compare its use by Christopher Wren about 1690, with the English Cross Bond of the Tudor Palace c.1530 (**Plate 26**).

Flemish Bond has for long been perceived as being weaker than English Bond but see Section 2.3.1 b.

Note: a wall more than one brick thick showing Flemish Bond on one side only is in **single bond**. If showing Flemish Bond on both sides it is in **double bond**. The latter might be required for a free-standing wall visible from both sides.

(d) Variations

Apart from Staggered Flemish Bond the other variations reduce the number of headers either by increasing the number of stretchers between headers in each course or by including one or more courses of stretchers between the single courses of Flemish Bond.

(i) Staggered Flemish Bond (Fig. 2.37)

The courses are identical to Flemish Bond but the overall appearance is quite different because the headers are positioned centrally over vertical joints in the course below rather than over stretchers. This produces vertical stripes of 'oscillating' headers. The quoin header is followed by another header instead of a queen closer. The alternate courses begin with three-quarter bats rather than stretchers.

(ii) Flemish Garden-Wall Bond (Fig. 2.38)

Also known as Sussex Bond, this bond has three stretchers between each header instead of one. The quoin header is followed by a queen closer and three stretchers. Alternate courses begin with two stretchers.

The bond, which is said to have been more popular than English Garden-Wall Bond, was sometimes used on the flank walls of houses that had Flemish Bond at the front.

Flemish Garden-Wall Bond has an advantage over Flemish Bond when building one-brick thick boundary and other

Fig. 2.36 Flemish Bond – with chequer pattern formed with dark headers.

Fig 2.37 Staggered Flemish Bond.

Fig. 2.38 Flemish Garden-Wall Bond.

freestanding walls, particularly if both sides are to be fairfaced. To achieve two fair faces in such walls the bricks must be selected so that only those that do not vary greatly in length are used as through-the-wall headers. There is no such advantage in this respect in using such bonds for a wall that is to have only one fairface, but that does not prevent its use for the sake of appearance.

It can be argued that walls built in this bond are weaker than those built in Flemish Bond because of the reduced number of headers or tie bricks. Although this is not known to have been confirmed by experience or tests and the structural masonry code of practice BS 5628: part 1 refers only to normal bonding, it would seem prudent to consult a structural engineer before building a heavily loaded wall in this bond.

(iii) Monk Bond (Fig. 2.39)

There are several versions of this bond which consists of two stretchers instead of one between the headers in each course.

In the first example the quoin headers are followed by two stretchers. The alternate courses begin with a stretcher and a queen closer producing a pattern similar in mood to Flemish Bond.

The second example contains four different courses (**Fig. 2.40**). The first and third courses both begin with a quoin header. That in the first course is followed immediately by another header and

Fig. 2.39 Monk Bond – headers rising vertically as in Flemish Bond.

Fig. 2.40 Monk Bond – a 'raking' version.

that in the third course by two stretchers. The second course begins with a three-quarter bat and a header and the fourth course with a three-quarter bat and a stretcher. This second version has a number of raking joints over nine courses which provide more 'movement' than the first example.

Monk Bond was used in medieval northern Germany and Scandinavia but not much in England. It is known as modified Flemish Bond in America.

Monk Bond was used in Guildford Cathedral (1936–61) by Sir Edward Maufe RA FRIBA. It was generally favoured by early 20th century Dutch architects such as Michel de Klerk and W H Dudok.

Fig. 2.41 Flemish
Stretcher Bond.

(iv) Flemish Stretcher Bond (Fig. 2.41)

This introduces, between single courses of Flemish Bond, one to six, but usually three, consecutive courses of stretchers each quarter-bonded with the one below and with the Flemish Bond course.

A quoin header and queen closer begin the first and third stretcher courses. A stretcher begins the middle stretcher course and the courses of Flemish Bond.

The three stretcher courses may be half-bonded and the queen closers omitted by beginning the first and second courses with a stretcher and the second with a header. The Flemish Bond courses begin with a quoin header followed by a three-quarter bat so that each header in the course is centrally over a stretcher.

In America the bond is sometimes called American Flemish.

e) Decorative bonds

(i) Herringbone Bonds (Fig. 2.42)

There are six main herringbone patterns. The direction in which the 'arrow heads' progress gives names to three basic forms, vertical, horizontal and diagonal and each of these may be single or double.

Vertical or Horizontal Herringbone Bond was often used in the past to infill early timber framed buildings possibly because the vertical and horizontal versions were easier to cut to an

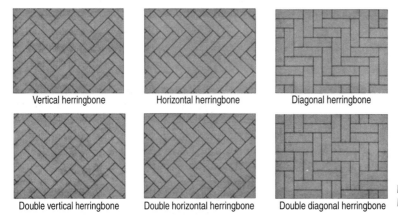

Vertical herringbone   Horizontal herringbone   Diagonal herringbone

Double vertical herringbone   Double horizontal herringbone   Double diagonal herringbone

Fig. 2.42 Types of Herringbone Bond.

Basket weave

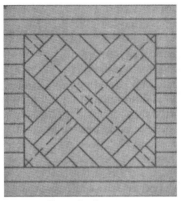

Diagonal basket weave

Alternative diagonal basket weave

Fig. 2.43 Types of Basket Weave Bond.

irregular space between timber framing than the horizontally bedded bricks of the diagonal version. It has also been suggested that as the newly laid diagonal bricks slump they fit tightly to help brace the frame, but this would be of value only if the faces of the frame in contact with the brickwork were notched to form a mechanical key for the mortar after the brickwork shrank on drying.

When used in facing brickwork today Herringbone Bond is usually formed as a panel surrounded by normal half- or quarter-bonded brickwork or between window openings.

Methods of setting out and building are illustrated in *The BDA Guide to Successful Brickwork*, Section 5.6 'Building decorative brickwork'.

(ii) Basket Weave Bond (Fig. 2.43)

Basket weave generally consists of three stretchers Stack Bonded with three 'soldier' bricks adjacent. Like Herringbone Bond it is usually formed as a panel set in an opening in normally bonded brickwork. The opening should be sized in multiples of brick sizes and built accurately. Diagonal basket weave can be set out so that the centre of the middle brick coincides with the centre of the panel or the main continuous joints can be set out to form an 'X' at the centre of the panel. Such a simple variation has a marked effect on the appearance of the panel.

(iii) Interlacing Bond (Fig. 2.44)

Sometimes referred to as 'pierced panelling' this bond has two versions, vertical and horizontal. Whole bricks are laid

Interlacing Bond

Diagonal Interlacing Bond

Fig. 2.44 Types of Interlacing Bonds.

vertically and horizontally, with one-third bricks to achieve the interlacing effect. During construction if one-third bricks are laid in sand they may be removed when the mortar has set to form a pierced panel. Alternatively one-third bricks of a contrasting colour may be built in.

f) Other bonds

(i) Stretcher Bond (Fig. 2.45)

This is most commonly seen in the half-brick-thick outer leaves of external cavity walls of brick buildings constructed since the 1930s and in virtually all post Second World War brick buildings. The cavity was introduced to improve the rain resistance of external walls and since the end of the 1970s they have been used to contain thermal insulation.

Each course consists of all stretchers most commonly half-bonded. Three-quarter bats are introduced to maintain bond at junctions and attached piers.

The bland neutrality of Stretcher half-bond is often appropriate in some contexts and, infrequently, quarter or third Stretcher Bonds have been used as variations but they can cause unsatisfactory broken bonds at quoins and window and door openings. Stretcher quarter-bond is often described as Raking Bond. Third

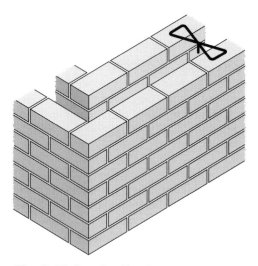

Fig. 2.45 Stretcher Bond.

bond was appropriately used for 290 by 90 mm modular bricks which have a stretcher to header ratio of 3:1.

The appearance of a quarter-bond such as Flemish, can be achieved by the use of half-bats or snap headers. If they are not supplied as 'specials' but are to be cut on site they should be preferably cut on a bench saw. See Section 3.2.2.

(ii) Header or Heading Bond (Fig. 2.46)

Consists virtually of all headers but with three-quarter bats beginning alternate courses. It has often been used to build curved brickwork as it is easier to follow

Fig. 2.46 Header or Heading Bond.

Fig. 2.47 Stack Bond.

a curve smoothly with headers only, than when stretchers are present. It was also used for whole elevations in some Georgian housing and other domestic buildings in the South of England, especially in Berkshire and Hampshire. This may have been because grey bricks were required and the bricks available had been set in kilns in such a way that only the headers were flashed. Header Bond is a very striking bond although it may appear a little harsh to some people.

(iii) Stack Bond (Fig. 2.47)

This describes brickwork in which several courses are 'stacked' one above the other with continuous vertical joints so that there is no overlap or bond. The bricks in each course may be bedded normally showing horizontal stretcher faces or on end in 'soldier' courses.

As excessive deviation in position or width of continuous vertical joints can be visually disturbing, it is advisable to have the bricks selected, preferably before dispatch to the site, in order to ensure that they do not vary greatly in length. Typically this might be no more than 2 to 3 mm.

The appearance of Stack Bond with its continuous vertical joints is more akin to tiling than brickwork and requires the practice of skills not acquired by all bricklayers. This is not to imply that it cannot and has not been used successfully given due care and attention.

(iv) Dearne's Bond (Fig. 2.48)

This bond consists of single courses of headers laid flat alternating with single courses of stretchers on edge with a space of approximately 75 mm between them within the wall. It was probably devised to provide low cost walls for humble buildings and freestanding walls.

Fig. 2.48 Dearne's Bond.

The bond is not generally suitable for facing brickwork as standard bricks are not faced on either the bed or upper surfaces, and many are perforated. However, it has been known for frogged bricks on edge to be used with the frogs exposed for decorative effect.

(v) Rat-trap Bond (Fig. 2.49)

Sometimes referred to as Silverlock's or Chinese Bond, this is based on Flemish Bond but all the bricks are laid on edge with the normal bed faces and header faces alternately visible. Alternate courses begin with two headers together.

It was used for low cost brickwork in, for example, cottages, almshouses and

Fig. 2.49 Rat-trap Bond.

freestanding walls and the less important façades of some buildings in which the main brickwork was in Flemish Bond. Its use has been noted in Bedfordshire and Hertfordshire.

Nowadays the bond would not be generally acceptable for facing brickwork as standard bricks are not faced on either the bed or upper surfaces. From the point of view of rain penetration, a wall of this bond should be considered as being a solid not a cavity wall.

(vi) Quetta Bond (Fig. 2.50)

Developed and first used by British Army engineers in Quetta, India in the 1920s to resist earthquakes, it is used to build walls one and a half bricks thick that can be reinforced vertically with steel bars. The face has the appearance of Flemish Bond.

(vii) Irregular Bond (Fig. 2.51, a, b, c)

Sometimes referred to as haphazard bond, this term is used to describe brickwork that has no regular bond as was common before brickwork began to develop as a facing material. The North Bar at Beverley, 1409–10, is an example

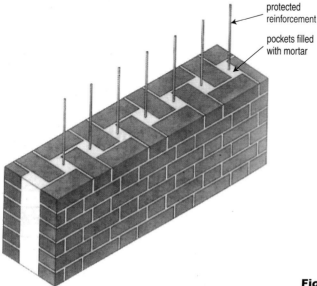

protected
reinforcement

pockets filled
with mortar

Fig. 2.50 Quetta Bond.

Fig. 2.51 (a) Irregular Bond at North Bar, Beverly, E.R. Yorks, c.1410.

more common in The Netherlands than it has ever been in England. Help!!!

Furthermore what the English call Flemish Bond is known in Germany as Polinischer Verband, or Polish Bond.

2.4 Enhanced brickwork

Plain, simple brickwork relying on colour, texture and bond pattern is often appropriate. But bricks, being varied in colour and texture and small with coordinated dimensions, can readily be used to enhance architectural features such as walls, window openings, quoins, plinths, string courses and copings. Bricks present sensitive designers with a broad and ample pallet to refine, enliven and enrich buildings and the spaces around them.

2.4.1 Surface patterns

a) With contrasting coloured bricks

Perhaps a medieval merchant or knight, home from trading or warring on the continent, began the trend for decorative brickwork in England as he inspected a delivery of bricks for his new manor house or castle. He may have noted the considerable number of flared or vitrified overburnt header faces inherent in bricks fired by unsophisticated means. The idea may have occured to him to make

of where bricks about $10\frac{1}{2} \times 5\frac{1}{4} \times 2$ inches have been laid with no discernible regular pattern.

Are you confused? It is hoped that this section has helped readers to gain a firm grasp of the many bond patterns. If any still feel confused the author can only sympathize, but we are in good company. Terence Smith currently Chairman of the British Brick Society has written

> ... it is clear that what the *English* call 'English Bond' the *Dutch* call 'Staand Verband' or 'Upright Bond'; the Dutch 'Engels Verband' ('English Bond') refers to what the English call 'Flemish Garden-Wall Bond or, alternatively, 'Sussex Bond', which is far

good use of such bricks by forming simple patterns in the manner he had seen abroad.

Diaper patterns were an obvious solution and an early English attempt was at the 15th century Herstmonceux Castle in East Sussex. Patterned brickwork was used more freely in the 16th century as at Layer Marney Hall, Little Leeze Priory and other buildings in Essex as well as the remaining portion of the Bishop's Palace at Hatfield. The commonly used English Bond was often broken for the sake of the diaper pattern as in an example in **Fig. 2.52** dated in the early 16th century.

In the earliest examples there seems to have been no attempt to select headers of the same colour and they often ranged through purples, blues and greys rather than blacks. The effect was subtle with the pattern sometimes almost dying away unlike such patterns made with consistent modern bricks (**Fig. 2.53**). As a more formal, classically inspired architecture began to replace the picturesque of the medieval and Tudor period so the use of diaper patterning declined, presumably being considered old fashioned and inappropriate for the new architecture.

When diaper and other patterns are used today, precision is probably appropriate and care is taken to plumb the patterns accurately.

Polychromatic brickwork, using several colours, was extensively used in early industrial buildings and in many types during the building boom of the Victorian era.

Many brick railway stations, hotels, civic buildings and churches were built with the self confident use of highly intricate and colourful decoration. Much of it was married with the Gothic revival greatly encouraged by the writings of John Ruskin and the work of architects such as Sir Giles Gilbert Scott, Pugin, William Butterfield and Samuel Sanders Teulon.

It was a period of great exuberance in the use of brickwork even if it was not to everyone's taste. However, there was a move away from this in the 1870s with a more eclectic approach to architecture bringing a revival of Jacobean and so-called Queen Anne styles owing much to the work of Sir Christopher Wren.

Right through to the years after the Second World War was a period first of austerity and then a preference for simplicity and under-statement. But during the 1970s, possibly as a reaction to the rather austere post-war buildings, there

Fig. 2.52 Diaper pattern in English Bond, early 16th century.

Fig. 2.53 Modern diaper pattern in Stretcher Bond, Wimbledon.

was a period when simple bands and more intricate patterned brickwork were used.

The bold simple forms used in the office building in Old Pye Street London SW1 might even have been stark were it not for the softening and interest of the diamond pattern using contrasting red bricks with yellow stocks. The building is of structural masonry and is in Flemish Bond (**Plate 28**).

b) Surface modelling

Brickwork surfaces can be modelled by projecting or recessing bricks beyond the face of a wall. It does not appear to have been used greatly until the 1920s onwards when it probably became more popular in order to relieve large areas of otherwise plain brickwork in, for instance, large blocks of flats (**Fig. 2.54**).

Projecting and recessed bricks may not be acceptable if they expose surfaces that are not faced, and in extreme cases frogs or perforations may be visible. Given early notice some manufacturers may be able to provide special versions of their standard bricks to overcome the problem.

Fig. 2.54 A brickwork surface modelled by projecting and recessing bricks.

Bricks used in this way should be frost resistant because accumulations of rainwater may saturate them in part. Furthermore, rain penetration of the brickwork may be increased, especially in exposed locations.

c) Coloured mortar joints

Patterns may be formed by the use of coloured mortars matching or contrasting with the bricks. It is seldom practicable to use more than one bedding mortar so that, in practice, the normal bedding mortar is raked out to a depth of 12 to 15 mm as jointing proceeds and subsequently pointed with coloured mortars. The apparent change in the colour of the brickwork is quite remarkable (**Plate 25**).

2.4.2 Articulation

When brickwork first became popular, coloured or moulded bricks were often used as dressings around window and door openings and for quoins, string courses and other features similar to those used in stone buildings. Bricks by their nature still lend themselves to the enhancement of modern buildings in many ways.

a) Band courses

These are a simple but effective way of giving a horizontal emphasis to a wall (**Fig. 2.55**). They may consist of a band of one or more courses contrasting boldly or subtly with the main wall in colour, texture or bond. They may be flush, project or be slightly recessed.

The concourse within the School of Engineering at De Montfort University, Leicester, 1993, has dark headers in Flemish Bond in the light-reflecting yellow calcium silicate structural spine wall and to the curved wall of an auditorium (**Plate 29**). This produces a precise crisp appearance and definite horizontal direction appropriate to an area in which students and staff move rapidly at

Fig. 2.55 Bold 3-course bands of darker bricks between four courses of light bricks at Hadrian Estate, Hackney, East London.

regular intervals. The banding echoes that in the external walls of clay bricks including the buttresses to the walls of the machine hall (**Plate 16**).

When bricks to be used in a band course are different from those in the main brickwork, perhaps by a different brickmaker, it advisable to check that the mean actual sizes of both will not vary by more than say 3 mm otherwise the variation in the widths of the cross-joints at the face may be visually unacceptable.

In band courses it is inadvisable to use bricks which have greatly different movement characteristics from those in the main wall as cracking may occur. Where two types of clay bricks are to be used the manufacturers should be consulted about their compatibility in the proposed circumstances. Clay bricks which expand over a long period of time should not generally be used with calcium silicate or concrete bricks which shrink.

A soldier course is simple yet attractive, generally consisting of bricks bedded on end showing stretcher faces. The bricks should be selected to have close tolerances about the mean length, typically ± 2 mm. Variations in the level of the top of the bricks will be particularly noticeable as will any brick that is not plumb.

Early consideration should be given to the method of returning the course at corners as some solutions require special shaped bricks. Six methods are shown in **Fig. 2.56 a–f.**

a) The standard special soldier brick continues the face dimensions of the soldier bricks round the corner.
b) A variant of a) echoes the half brick dimension in the courses above and below but shows a continuous vertical joint on one face.
c) Uses a standard special that is faced on one bed.
d) Uses standard bricks, Stack Bonded but gives an asymmetric corner.
e) A bonded corner of standard bricks is better balanced than d).
f) Uses a purpose-made return brick, 140 mm square with two false joints giving the same appearance as a) but is more robust.

The presence of soldier courses, particularly if multiple, may reduce the resistance of the wall to both compressive and lateral loads, in which case structural engineering advice should be sought.

b) Dentilation

Dentilation or dentil courses are a traditional form inspired by classical cornices to articulate the top of a wall. They consist of a regular pattern of projecting headers spaced centrally over stretchers in the course below. The course above the dentil consists of stretchers placed centrally over and flush with the faces of the headers (**Fig. 2.57**). See also Reading Town Hall (**Plate 30**).

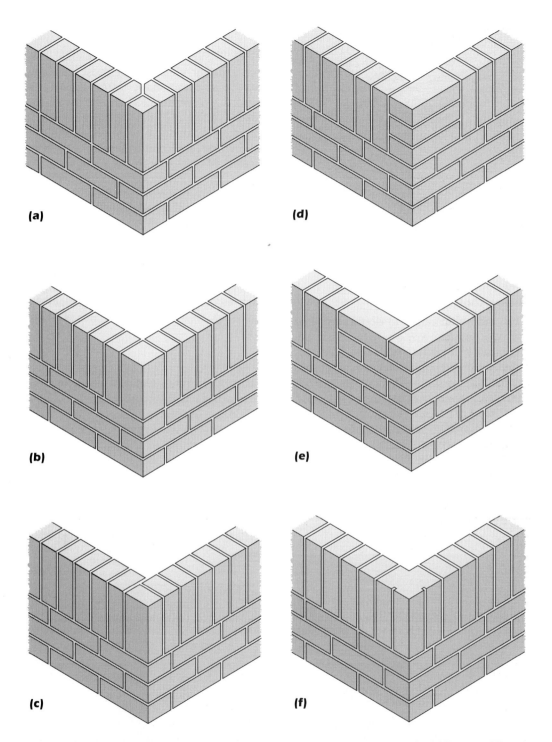

Fig. 2.56 Alternative returns to soldier courses at external angles; (a) standard 65 mm × 65 mm 'soldier return'; (b) standard 102 mm × 102 mm 'soldier return'; (c) standard 102 mm × 65 mm 'soldier return'; (d) standard bricks, Stack Bonded; (e) bonded corner using standard bricks; (f) purpose-made return brick.

Fig. 2.57 Dentilation at a verge.

c) Dog toothing

This is an alternative to dentilation having two basic forms, one projecting and the other recessed (**Fig. 2.58**).

In both forms, two faces are set at 45° to the line of the wall face. In the former, each arris between two faces projects from the face of the wall and in the latter the arrises are flush with the wall forming a recessed feature. Great care has to be taken in setting out, aligning and levelling. The method is set out in some detail in *The BDA Guide to Successful Brickwork*, Section 5.6 'Building decorative brickwork'.

d) Dressings

i) Door and window openings

Being major elements of most masonry buildings these have often been enhanced by the detailed treatment of the heads, jambs and sills. From the earliest days of brickwork bricks were moulded to provide architectural detailing reminiscent of carved stone dressings. Many of the moulded brick features were in fact rendered to imitate stone dressings. From the early 17th century, brick versions of classical details in moulded or gauged brickwork were fashionable.

Today the cost of labour and architectural fashion tend to restrict such enhancement to the use of special shapes such as cants and bullnose to model the jambs as well as sills and heads (**Fig. 2.59**).

ii) Quoins

These were often built using a form of rustication similar to that used in stone buildings. But a simpler form of enhancement is to create block bonded corners, possibly in contrasting colours (**Fig. 2.60**).

Fig. 2.58 Dog toothing under an eaves.

Fig. 2.59 Specials used as a modern dressing to a window opening.

Fig. 2.60 A rusticated brickwork quoin.

2.4.3 Brickwork features

Well detailed building elements often make a major contribution to the appearance of a building. A number of common brickwork elements are considered below.

a) Curved brickwork

Brickwork may be curved in plan, purely as an aesthetic feature or for structural advantages in resisting lateral and compressive forces. It is advisable to consider whether it is possible to use standard straight bricks or whether radial, i.e. curved, bricks are advisable.

(i) Standard straight bricks

These may be used to build brickwork to gentle curves. But the smaller the radius of a curve the greater will be the amount that the straight faces of adjacent bricks are out of line, the 'faceting' effect, and the greater will be the amount each brick in one course overhangs the bricks in the course below (**Fig. 2.61**).

The extent to which these factors are acceptable depends not only on the radius but also on the type of bond. The overhang will be most noticeable in Stretcher half-bond, and less noticeable with quarter-bonds such as Flemish and English, and least with Header Bond (**Fig. 2.62**).

Fig. 2.61 The effects of 'faceting' and 'overhang' accentuated in strong oblique light.

Strong directional lighting will of course emphasize any discrepancy (**Plates 31 and 32**).

The smaller the radius of curvature the more the face width of vertical cross-joints will diverge from the nominal 10 mm in straight brickwork, but this may not be unacceptable. The point is illustrated by the half-brick thick serpentine wall which has a radius of 3.25 m. The average widths of the cross-joints are 8.5 mm on the concave face and 15 mm on the convex face yet, surprisingly, the difference is hardly noticeable and does not affect the overall apparent consistency of the brickwork (**Fig. 2.63**).

Design guidance in tabular form on the use of straight bricks is given in the Brick Development Association's Design Guide 12 'The Design of Curved Brickwork'. The tables cover half-brick and one-brick walls in stretcher, Flemish and Header Bond, for four selected radii and the cross-joint widths for both convex and concave sides. Formulae are given to calculate the joint widths for particular radii not included in the tables.

Because of the divergence in joint

Fig. 2.62 A curved wall of headers.

Fig. 2.63 A serpentine wall of half-brick thickness.

width, already explained, variations in the work sizes of standard straight bricks are more significant in curved brickwork than in straight walls. It is advisable with small jobs to use bricks from one consignment only or, with large jobs, to make special arrangements with the supplier to minimize the dimensional variation.

(ii) Radial bricks

Radial bricks with the same radius as the brickwork give a more precise and refined appearance than can be obtained with standard straight bricks (**Fig. 2.64**). Radial bricks can be purpose-made to order for any radius with either the convex or concave surfaces faced.

However, a range of six standard radial stretchers and six standard radial headers is specified in BS 4729. They are provided for building curved walls with a convex fairface of six different radii, ranging from 450 mm to 5400 mm, and measured to the outer, convex face.

Much curved brickwork is in the form of a quadrant at what would otherwise be a right angled corner. Each of the radii prescribed in the standard is a multiple of 225 mm, the coordinating dimension of a standard brick, to enable a neat bond to be made between the ends of the quadrants and the straight walls. The use of purpose-made 'transitional' bricks, having one half their length straight and the remainder curved, will give a smoother transition from the curved to the straight wall if the bonding is to continue. Alternatively a straight joint might be considered.

b) Copings, cappings and sills

Properly designed and constructed copings and cappings not only keep the brickwork which they protect from staining and eventually deteriorating, they also provide an architectural finish to freestanding, retaining and parapet walls and chimneys. In this respect

Fig. 2.64 Radial stretchers for a small radius.

Fig. 2.65 A double cant-on-edge capping with an external return.

Fig. 2.66 A brick saddleback coping.

Fig. 2.67 A cast stone coping provides a prominent architectural finish to a parapet wall.

they may range from minimal brick-on-flat and brick-on-edge flush cappings – either square, or with bullnose or cant profiles – through simple saddle back copings to prominent, weighty projecting copings (**Figs. 2.65–2.67**).

Copings have a continuous groove or 'throat' along the projecting underside to prevent run-off rainwater tracking back to the face of the wall and eventually flowing down and wetting it. It is essential that the throat is kept clear of mortar during construction particularly at the joints between the ends of coping units.

A capping has no throat and is usually finished flush with the wall. Flush detailing was much favoured during the 1970s and 1980s but in recent years a marked preference for copings, particularly in stone, has developed as designers have rediscovered how attractive different but complementary materials for adjacent elements can be.

Flush, brick cappings can be formed from standard bricks on edge or standard special bullnose or cant bricks for alternative profiles. If a brick capping to a one-brick thick wall is to be fairfaced both sides, it is advisable to select the bricks in order to discard over-long or short bricks that will not align both sides. A full range of standard stop ends and external and internal returns are available to give a neat and robust finish.

Because flush details allow rainwater to run-off directly over the wall below, the brickwork must not only be designed to resist the common results of saturation such as frost attack, see Section 4.3, but the possibility of staining should be considered. Flush detailing can be satisfactory but requires particularly careful design and construction.

Standard copings and cappings are described in Section 2.1.1 g(iii).

Similarly, sills range from those that minimize run-off over the wall below by providing stooling at the ends as well as a drip through to a flush detail (**Figs. 2.68–2.70**). These not only have very different appearances initially but may weather quite differently as described in Chapter 4.

c) Arches

The arch form provides a structurally elegant way of transferring the vertical forces from above an opening to abutments either side, and must be one of the most common and readily recognized elements of masonry construction. Arch terminology is illustrated in **Fig. 2.71**.

The satisfying appearance of arches probably derives from their evident structural purpose and the aptness of the tapered voussoirs which wedge together when under load. The mortar joints serve only to distribute the load uniformly, not to provide adhesion. Seldom has the wisecrack that mortar keeps bricks apart rather than holds them

Fig. 2.68 Projecting and stooled stone sill.

Fig. 2.69 Projecting sill with purpose-made bricks.

Fig. 2.70 Flush sill with purpose-made bricks.

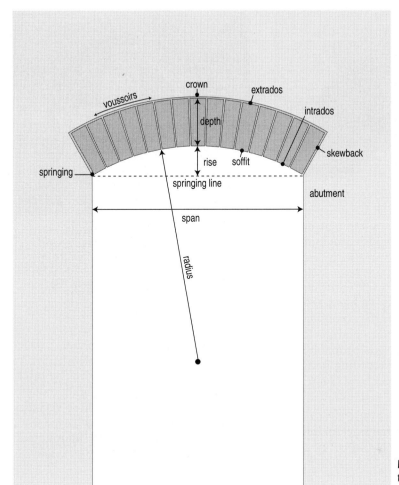

Fig. 2.71 Arch terminology.

together been more apposite. The colour, texture and human scale of brick voussoirs serve to increase the attraction.

In Britain we are accustomed to the prolific use of arches particularly in 19th century bridges, warehouses, factories and housing of all types. However, in most modern brick buildings, the roof and floor loads are usually carried by a structural frame or, in the case of domestic buildings, by the inner leaves of cavity walls so that the outer leaf of brickwork carries only its own weight and arches may be structurally unnecessary.

But arches have been a feature of brickwork for so long that many people are uncomfortable with say a course of stretchers over an opening with virtually no visible means of support. Consequently, many designers have begun to use a true arch form or a symbolic substitute even if there is no structural need. At the same time a small but growing number of architects and engineers interested in using an appropriate technology are beginning to use structural brickwork.

The main types of arches, both past and present, may be conveniently considered under five types.

(i) Rough arches (Fig 2.72)

These are built from standard bricks with wedge-shaped joints between. They are

Fig. 2.72 *A semi-circular rough arch.*

Fig. 2.73 *A double semi-circular arch of voussoirs delivered to site cut to shape.*

usually confined to semi-circular or seg-mental arches and are probably not only acceptable but apposite with rugged, stock-type bricks. Smooth, even-coloured bricks seldom look well with tapered joints. In rough arches of headers span-ning more than 1.8 m the small taper to joints will often be acceptable. Rough arches of stretchers are probably only acceptable for spans exceeding 3 m.

(ii) Axed arches

Traditionally, voussoirs were cut to shape on site with a lump hammer or bolster, trimmed with a scutch and rubbed with a carborundum block.

(iii) Using standard arch bricks

Section 2.1.1g(iii) describes standard special arch bricks, consisting of tapered headers and stretchers suitable for circu-lar arches of specific 'ideal' spans.

If a substantial number of tapered bricks are required to build arches which are not one of the 'ideal' spans it might be preferable to use purpose-made non-standard arch bricks.

(iv) Preformed arch sets

Voussoirs for axed arches are seldom cut on site today. Many manufacturers will prepare computer aided drawings and cal-culate the sizes of voussoirs for particular arches and cut them to shape before deliv-ery in sets. They often use special shaped

units such as bullnose and cants to model the intrados (**Plate 33**) (**Fig. 2.73**)

(v) Gauged arches

Guaged arches were traditionally built from bricks known as 'rubbers' which were made from fine clays blended with a high percentage of fine sand. The bricks were consequently soft enough to cut and rub to shape and size on site. Very fine joints of 1 mm or less were achieved using lime putty. Gauged arches undoubtedly give a refinement and sense of quality to brickwork (**Fig. 2.74**).

Fig. 2.74 *A segmental gauged arch.*

A few manufacturers make traditional 'rubbers' but they are somewhat costly and are used mainly for high quality restoration work. Other manufacturers offer bricks of a similar appearance, mechanically formed to shape before delivery. Joints 2–3 mm thick using lime putty can be achieved depending on the type of brick, but with cement: sand mortars 6–10 mm joints would be more usual.

It is not possible in this book to illustrate the full range of arch forms which have been used in brickwork but **Fig. 2.75** is an example of a variation of a simple cambered arch while **Fig. 2.76** is an example of the more elaborate forms.

Fig. 2.75 A cambered Welsh arch.

Fig. 2.76 Gothic arches with decorative voussoirs and keystones.

d) Corbels and projections

Projections of continuous or isolated features capable of supporting loads have been readily built by the considerable ingenuity exercised in the past by brick-builders. Such features have for long been a characteristic of brickwork (**Plate 30**).

Corbels and other projections have frequently been used for practical purposes or as architectural features such as projecting eaves, supporting external right angle corners over splayed corners or supporting splays over internal angles. Corbels were a common feature in domestic buildings traditionally found at verges and eaves often with dentil courses.

The neat corbel feature using inverted purpose-made plinth bricks to make the transition from a splayed reveal abutment on which the skewback rests will be a delight to anyone with some understanding and appreciation of brickwork (**Fig. 2.77**).

Similarly, in recent years, many buildings use a visual transition from a plain wall to a projecting feature above (**Fig. 2.78**). They are not true corbels, there being unsufficient thickness and weight of brickwork to be so and they are supported by a sophisticated stainless steel

Fig. 2.77 Corbelling from a sprayed reveal to a right angled abutment above.

Fig. 2.78 *'Special' plinth bricks inverted to form 'corbelled' feature beneath a projection.*

system fixed to the building's structural frame. This may offend the purist but otherwise we may appreciate the nod towards a familiar past achievement.

It should be noted that, if the projection of one brick over another is excessive, the result may be coarse and clumsy. Reducing this amount can add to the refinement of the feature. Note must be taken that bed and stretcher surfaces that are not faced may be exposed and the former may contain frogs or perforations. Brickmakers can usually provide purpose-made special bricks to deal with such situations.

e) Plinths and sloping surfaces

When solid, external brick walls are reduced in thickness at some level, by stepping the external face back by a half brick, the risk of saturation from rain collecting on the resulting ledge and damage to the arrises can be elegantly avoided by the use of plinth bricks. See Section 2.1.1g(iii). The bold use of multi-coloured plinth bricks in a battered wall is entirely appropriate as part of the functional nature of the River Foss flood alleviation scheme in York (**Plate 34**).

Plinth bricks have a 45 degree chamfer and are dimensioned so that two courses are set back by the required half brick increment. As a result the vertical surface below the chamfer is only 9 mm high. This small size can be difficult to form precisely with soft-mud and hand-made bricks and may be liable to damage during handling. It can also lead to distortion when forming bricks by extrusion as any slight dimensional variation is often noticeable in such a small upstand. Fortunately, an alternative plinth brick, dimensioned to set back by a half-brick increment in three courses, has a more robust 23 mm upstand.

In case of doubt concerning the suitability of the 9 mm upstand for particular types of bricks, consultation with the manufacturer is recommended.

The 9 mm high face gives an overall slope to the plinth of about 61 degrees while the 23 mm face gives a slope of about 53 degrees.

Sloping surfaces can be formed by multiple courses (**Fig 2.79**) but because of the differing girth at corners some closers are necessary which have to be cut from appropriate plinth bricks as standard closers are not specified in the British Standard.

Fig. 2.79 *A plinth of standard plinth bricks.*

Fig. 2.80 Vertical movement joints in a clay brickwork outer leaf.

f) Movement joints

Movement joints built into brickwork to allow movement without cracking inevitably affect the appearance of the building unless steps are taken to hide them from sight. The visual impact will depend on where they are placed and how they are detailed and built (**Plate 35**) (**Figs. 2.80–2.82**).

All external brickwork continually expands and contracts as it is alternately exposed to and shielded from sunlight, and as the air temperature rises and falls. In addition, clay bricks undergo irreversible long term moisture expansion, while calcium silicate and concrete bricks shrink irreversibly. Depending on the geometric form of the brickwork and the degree of restraint offered to potential movement the brickwork may crack or buckle.

To prevent either of these happening the design and position of movement control joints should be considered early in the design and not treated as an afterthought when the main structural decisions have been made. They may be either detailed as a feature or placed unobtrusively.

Movement joints are essential in most half-brick thick leaves of modern buildings as they are relatively lightly loaded and unrestrained. In many older brick buildings the greater self-weight of the

Fig. 2.81 A feature made of a vertical movement joint with cant bricks.

Fig. 2.82 A vertical movement joint is less prominent in an internal corner.

brickwork and the loads from floors and roofs offer more resistance to movement. Futhermore lime mortars permit movement to be taken up in small increments within the joints rather than be concentrated in one place as tends to happen in brickwork built with cement mortars.

Movement joints may be honestly expressed as vertical lines, in which case an appropriately coloured sealant is important and some expertise is required to finish it neatly. Of course the design and workmanship must primarily ensure that the movement joint fulfils its function. This is dealt with in greater detail in *The BDA Guide to Successful Brickwork* Section 4.5 'Building movement joints'.

g) Chimneys

Chimneys are more than devices for conducting hot smoke and fumes from fireplaces and furnaces to the open air. They are also visual symbols of human achievements, whether they be large or small, serving dwellings, factories or power stations.

We see around us a variety of chimneys. Tall, Tudor and Jacobean chimneys often elaborately decorated with cut and moulded bricks demonstrated that the owner lived in a modern house with many fireplaces and was very rich (**Plate 5**).

With the coming of the classically inspired architecture of the 17th and 18th centuries, chimney stacks became simpler in form and less decorated but, during the 19th century, chimney design again became more elaborate. Tall stacks came into use from the late 18th century in connection with mining and steam engines providing power for mills but they were mainly built in stone.

During the 19th century many were built in brickwork, and increasingly those serving mills became freestanding rather than part of a building. The motive to build them high and decorated probably stemmed from a desire to advertise and impress. Polychromatic designs often incorporated the name of the mill.

Brickwork ventilation stacks terminating high above the roofline of the Engineering School at De Montfort University, Leicester, 1993, demonstrate the visual impact of such structures which in this case also buttress the stuctural brick walls within the building as chimneys attached to buildings often do, whether by design or fortuitously (**Fig. 2.83**).

Designers today still take the opportunity to enhance a chimney as at Queen Mary's Hospital, Hampstead. Here, the simple vertical recesses and the carefully

Fig. 2.83 *Tall stacks ventilate a School of Engineering.*

proportioned projecting bands and corbelling at the terminal turn a utilitarian feature into one of elegance (**Plate 36**).

But as chimneys are one of the most exposed of building elements, care is needed in the design of the terminal in the interests of durability as well as appearance. Those built from half-brick enclosures to flue liners require extra care in the design and construction of damp-proof course trays and flashings.

2.4.4 Gauged brickwork

Gauged brickwork is laid to a fine degree of accuracy using special bricks called 'rubbers' made of blended washed clays and sands and fired to allow for rubbing and cutting accurately to shape and size; hence the original term 'cut and rubbed work'. The bricks were usually laid by dipping into a slurry of slaked lime putty and silver sand to form joints not exceeding 2 mm thick (**Plate 37**) (**Figs. 2.84 and 2.85**).

Gauged brickwork emphasizes the whole mass rather than the individual bricks and the patterns they make and was ideal for creating dressings of arches,

Fig. 2.85 A segmental gauged arch.

aprons, cornices and pilasters with carved *in situ* capitals using soft red, orange and yellow rubbers which contrasted with the general walling in colour as well as the use of thin joints.

The 'Dutch House' or Kew Palace, c.1631 (**Plate 6**), is believed to be the earliest example of rudimentary gauged work as is Cromwell House, Highgate, North London, c.1637. By the middle of the 17th century the art of gauged brickwork had reached a degree of perfection, as can be seen in number 3, King's Bench Walk, London, c.1667 by Sir Christopher Wren.

Gauged work enjoyed a revival in popularity in the second half of the Victorian period with the 'arts and crafts' movement and remained popular until the period immediately following the First World War when it declined in popularity due to the lack of patronage by wealthy clients and the loss of highly skilled craftsmen during the war.

2.5 Brickwork with other materials

Many fine buildings exist in which the walls are entirely of brickwork but there is no denying that the charm of many stems from the juxtaposition of brick and other materials.

An early instance was the use of brick

Fig. 2.84 A gauged brick niche.

dressings at quoins and window and door openings in walls of flint or stone rubble because, it may be reasonably assumed, it was easier and quicker to build them accurately, neatly and robustly in bricks than in the more irregular materials that could not be bonded but were local and more plentiful than brick.

What began as a functional solution may have led to an appreciation of the aesthetic possibilities. The use of gauged brickwork as dressings, described in Section 2.4.4 above, was often used in walls of other types of brick or stone. The early instance was reversed when the dressings were in ashlar with the main wall in brick as at Tattershall and Hatfield. But it must not be supposed that this was always done for economy, for it is apparent that the wealthy patrons of Hampton Court Palace appreciated the colour value of brick associated with stone.

Red brick was widely used with white paint as at Wren's Morden College, Blackheath, 1695, where quoins, bands and columns are of brick plastered and painted white – the cornice is of wood painted white.

In 1992 Mercers' House for the Mercers' Company Housing Association, in Essex Road, North London, successfully combined 50 mm high bricks gauged at five courses to 300 mm, and ranging in colour from brown through orange to pink with cornices, copings, string cour-

ses and sills in concrete with a crushed stone aggregate (**Plate 38**).

Richmond House, a new office block in Whitehall close to Inigo Jones' Banqueting House, called for sensitive treatment. The solution is said to have been inspired by the brick buildings of the first Elizabethan age (**Plate 39**).

2.6 Hard landscaping

2.6.1 Freestanding brick walls (Figs. 2.86–2.88)

Freestanding brick walls, whether they are defining the boundaries of a site, dividing it, providing privacy, security or noise barriers make such an important contribution to the quality of private and public spaces that they should be thoughtfully designed and built with care.

They must be designed to resist severe gusts of wind for, when walls are blown over, investigations invariably reveal that they have been merely put up rather than designed.

The means taken to ensure stability can often enhance the appearance. Piers are structurally most effective if they project equally on either side of a wall but it may be visually preferable if they appear on one side only when they will have to be designed accordingly. The piers may be the same height as the wall or higher or lower. Small piers projecting a little above the main wall can look very fussy.

Fig. 2.86 A serpentine stock brick wall is a perfect foil for soft landscaping.

Fig. 2.87 A staggered wall provides stability and shelter for planting.

The type of coping or capping used is most important in determining the appearance.

Alternatively, if space permits, the wall may be staggered in plan or curved to give it inherent stability and attractive spaces for planting.

As the top and both sides of a free-standing wall are exposed to wind-driven rain the brickwork can become saturated and liable to staining and possibly physical deterioration. A properly designed and appropriate coping will not only help protect the wall but can visually enhance it. It may be minimal or bold enough to accentuate the top of the wall.

As most freestanding walls will be at least one brick thick they provide an opportunity to use one of the quarter-bonds described in Section 2.3.2. If they are to be fairfaced both sides, it may be necessary to select bricks for the through the wall headers for size consistency so that it will not be too difficult to keep two fair faces. This is where Garden-Wall and similar bonds which reduce the number of headers through the wall may be appropriate.

2.6.2 Earth-retaining walls (Plate 40) (Figs. 2.89–2.91)

The appearance of sloping ground in the landscape can often be dramatically improved by sudden changes of level which also provide extra useful flat spaces. The vertical faces are greatly enhanced by brick earth-retaining walls designed to resist the lateral thrust and so contain the earth (**Plate 40**). For these walls to retain their appearance the back of the wall in contact with the earth must be protected by a damp-proof membrane to prevent water continually seeping through the brickwork giving rise to efflorescence, lime leaching, the growth of lichen or, worse, frost attack or sulphate attack of the mortar. See Section 4.3 and **Fig. 2.92**.

Fig. 2.88 A boldly patterned wall can enliven a dull urban scene.

Fig. 2.89 A curved brick diaphragm earth-retaining wall.

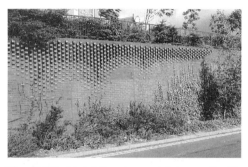

Fig. 2.90 A brick-faced r.c. retaining wall with a pattern of projecting headers.

2.6.3 Paving

Clay bricks have been used for surfacing pavements for vehicles and pedestrians since medieval times. In recent years the development of purpose-made clay pavers and the establishment of the flexible form of construction has led to their wide application especially where appearance has been important.

Calcium silicate bricks are not recommended for use as pavers where de-icing salts are liable to be in contact with them.

The small size of paving units permits a wide variety of intricacy and form appropriate to the size and context of the design.

a) Flexible paving (Plates 41–46)

Clay pavers laid on a bedding course of sharp sand on a granular base have a proven performance in service in European projects completed in some cases during the 19th century. Their construction was based on experience and 'rule of thumb' techniques.

The advantages of flexible clay pavements were recognized by UK clay brick manufacturers in the late 1970s. Since then and in conjunction with the Brick Development Association they have been involved with promotional activities and extensive laying trials and have sponsored research to establish guidance for designers as well as practi-

Fig. 2.91 A retaining wall provides a flat approach to the dwellings above and parking space below in a road widening scheme.

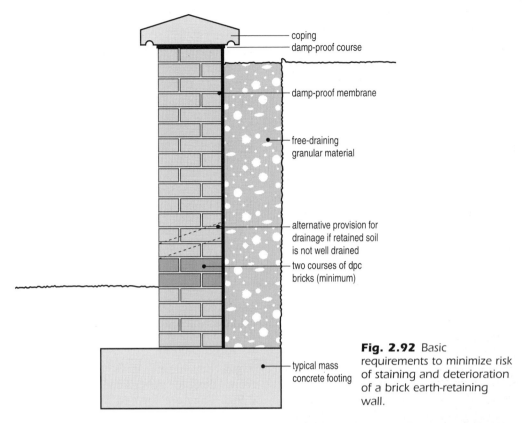

coping
damp-proof course

damp-proof membrane

free-draining
granular material

alternative provision for
drainage if retained soil
is not well drained

two courses of dpc
bricks (minimum)

typical mass
concrete footing

Fig. 2.92 Basic requirements to minimize risk of staining and deterioration of a brick earth-retaining wall.

cal recommendations for laying flexible pavements.

Flexible paving has been used in a wide range of applications from domestic patios and driveways, residential estate roads, major town-centre pedestrianization schemes, hard landscaping and prestigious public and commercial developments to heavy duty use such as bus stations and industrial sites. Many successful schemes have also been completed, encompassing the full spectrum of pedestrian and vehicular loadings. Flexible pavements are now an accepted and economic alternative to rigid pavements. See Section (b) below. Clay pavers are preferred on schemes where appearance is considered particularly important (**Fig. 2.93**).

Flexible pavements in public areas should preferably be designed with the assistance of experienced highway engineers and be constructed by specialist

Fig. 2.93 Flexible paving in Herringbone Bond.

paving contractors, especially if the design requires intricate cutting and fitting. Guidance on the design and construction of flexible pavements is available in the form of British Standards or from publications produced by the

Brick Development Association or paver manufacturers.

i) Bond patterns

The recommended laying pattern for pavements subject to regular vehicular trafficking is Herringbone Bond as it maximizes interlock between pavers in the wearing surface to resist the acceleration and braking forces imposed by vehicles (**Plate 44**).

The most common size of paver used is one with a 2:1 bonding ratio, e.g. 200 × 100 mm. Other bonding ratios such as 3:1 or pavers with brick size dimensions may also be used dependent on the level of trafficking envisaged.

For pedestrian areas and driveways to private houses a Running Bond pattern may be used. It is particularly suitable where 215 × 102.5 mm pavers are to be used. Running Bond may not be suitable for small radius curves as joints between pavers taper to an unacceptable degree and it is impossible to maintain the bond pattern without frequent use of batts.

It is essential to ensure that a suitably robust edge restraint is provided to prevent the lateral spread of the pavers and the opening of the joints. Edge restraint may be provided by existing features such as walls or by purpose-made kerb or channel details.

ii) Surface drainage

Surface gradients and the positioning of drainage channels and gully outlets should of course be carefully considered to avoid ponding, but they are also an important part of the appearance of the pavement which can be such a conspicuous element in the built landscape (**Fig. 2.94**).

Drainage channels may be formed from rectangular pavers or from purpose-made channel units from some manufacturers.

Falls to point outlets are difficult to form in flexible clay pavements and should be avoided if possible. Where

Fig. 2.94 *A surface drainage channel in flexible paving.*

there are several such points a disturbing wavy effect may be created over the whole area.

The sand-filled joints between pavers in the wearing surface of a flexible clay pavement will gradually seal with time and become effectively impervious.

b) Rigid paving

Rigid pavements look quite different from flexible pavements (**Figs. 2.95-2.96**). They are formed by laying pavers on a mortar bedding course over a rigid foundation and fully filling the joints between with mortar. Joints in rigid

Fig. 2.95 *Rigid paving, Herringbone and Stretcher Bond with movement joints.*

Fig. 2.96 An attractive landscape feature in rigid paving.

pavements are typically 10 mm wide. The method requires more skill and is slower and consequently more expensive than the flexible method. Rigid pavements are suitable for use in pedestrian and lightly trafficked areas.

c) Movement joints

Movement joints are not required in flexible pavements as thermal and moisture movements are accommodated within the joints. However, movement joints are advisable at about 6 m intervals if clay kerbs, channels or pavers are used to form rigid edge restraints. With clay pavers, movement joints are necessary at a maximum of 6 m within the paved area and at the margins.

Movement joints should be formed through the rigid paving including the mortar on which it is bedded, and then filled with an eminently compressible filler such as polyurethane foam and made weather proof with a tack-free, non-setting sealant. Movement joints can often become part of the overall design if considered in conjunction with the bonding and can be emphasized by using pavers of a contrasting colour adjacent to them.

2.7 Carved brickwork

2.7.1 Early examples of carved brickwork

A Persian archer or royal guardsman was the subject of one of the earliest examples of brick sculpture on a wall of the palace of Darius at Susa. The figure which stands some 17 courses high was built up from about 50 coloured glazed bricks about 500 to 350 BC. It is at present on display in the British Museum.

In England the carving of stone and brickwork was practised from early times. At East Barsham Manor House, Norfolk, the coat of arms of Henry VII over the entrance to the house is of carved stone whereas the coat of arms of Henry VIII over the gatehouse is of carved brickwork. During the 17th and 18th centuries much refined architectural embellishment consisted of carved soft brickwork with fine joints.

2.7.2 Modern brick sculpture

Sculpture often conjures up visions of monumental figures carved in stone or fine marbles, and perhaps to many, brickwork seems too commonplace to be worthy of the attentions of an artist. Yet Walter Ritchie, a versatile artist carving a diversity of materials including marble, limewood, sycamore, perspex and steel as well as making jewels in ivory, silver, gold, rock crystal and malachite has created many diverse and uncommon carvings in 'commonplace' brickwork.

Techniques for producing sculptured brickwork include:

- carving the brickwork *in situ*
- carving prepared brick panels to be transported and fixed on site later
- carving bricks in the green state which are then fired and assembled on site.

Walter Ritchie has said of his brickwork sculpture that much '. . . is "architectural" sculpture which may be small and intimate and give surprise in an unexpected corner of a building, or may be a significant part of an elevation'. Much of his brick carving is in low relief; first, because that has been appropriate in the circumstances and second, because it is less expensive than work in the round.

Fig. 2.97 Part of 'The Creation', a brick sculpture at the Bristol Eye Hospital.

Fig. 2.98 A sculpture in Cambridge.

'The Creation' at the eye hospital in Bristol came about because the planning authority approved the design on condition that some elevational enrichment was added at street level. Not wishing to introduce another material into an otherwise red brick building the architect asked Walter Ritchie to design sculpture panels in brickwork. The panels took him three years to carve in his studio at Kenilworth using bricks with a matching coloured mortar (**Plate 47**) (**Fig. 2.97**).

Lettering carved large on a bridge in Taunton, Somerset, will surprise and delight many. Surprise, because we are so conditioned to thinking of lettering being carved into marble, granite or fine stone, that it is surprising to see bold, large lettering carved in brickwork such as that by Richard Kindersley the sculptor (**Plate 48**). The delight is in the way the top and bottom of the whole inscription respects the straight horizontal bed joints but the rest of the lettering flows freely over the brickwork echoing the movement of waves on water. The letters have been edged in by chisel and carved across as a cushion section within

which the recessed joints have been filled. The complete carved section was then tinted with a brick dye.

The brickwork is in stretcher half-bond and coloured mortar matches that of the bricks but the bond pattern is readily apparent as the joints have been recessed so that the brick edges cast a shadow. Richard Kindersley has carried out lettering and graphics at St Paul's Cathedral, Westminster Abbey and the Queen Elizabeth Bridge at Dartford.

Fig. 2.99 A figure of Justice at the Magistrates' Court, Wigan.

Fig. 2.100 Coat of Arms.

David Kindersley produced a carving on a building in Cambridge which has a surprising delicacy considering the heav- iness of the brickwork from which it is carved (**Fig. 2.98**).

Internal low relief carvings enliven what may otherwise have been an un- eventful entrance corridor in the Council offices in Morpeth, Northumberland. They were carved by John Rothwell, an architect with no previous experience of carving. The bricks are hand-made with typical creased faces (**Plate 14**).

Unfired 'green' bricks assembled as a panel were carved by Christine Ward for a brickwork sculpture depicting 'Justice' in the Magistrates' Court in Wigan (**Fig. 2.99**). After carving, the panel of bricks was dismantled and each brick num- bered before drying and firing. They were then built on site. Another carving of the Coat of Arms was also made for the same building (**Fig. 2.100**).

3 Site practices as they affect the appearance of brickwork

3.1 Site management – good practice begins before a trowel is lifted

To produce high quality facing brickwork, bricklayers need the support of knowledgeable far-sighted site management. Important management tasks include:

- early contact with brick suppliers
- early ordering
- reasonably accurate estimating
- scheduling of deliveries, especially of special shaped bricks
- planning or beginning the following matters before the first materials are delivered and involving the foreman bricklayer or team leader in these early preparations.

3.1.1 Checking deliveries of facing bricks for minor blemishes

Deliveries of facing bricks inevitably include a few with minor blemishes. It is not possible to define an upper limit at which rejection of the delivery would be justified, not least because it has not proved possible to devise a standard test for compliance.

The British Standard for clay bricks states that, 'As a guide, bricks should be reasonably free from deep or extensive cracks and from damage to edges and corners, from pebbles and from expansive particles of lime.' But because in practice it is difficult for the supplier and customer to agree what is reasonable, Appendix F of the Standard gives detailed guidance on a method of agreeing a 'reasonable' and acceptable level of minor blemishes. This involves building and comparing 'reference' and 'sample' panels. The guidance is summarized below.

a) Reference panels (Fig. 3.1)

The purpose of a reference panel is to provide a basis for judging the accept-

Fig. 3.1 A reference panel.

ability of minor blemishes on the faces of future deliveries of bricks. A satisfactory judgement cannot be made by examining individual bricks.

Before bricklaying begins, a reference panel should be built showing at least 100 brick faces. These should have been either delivered by the supplier as being representative of the average quality of future deliveries or they should be randomly sampled from deliveries.

The panel should be built to a standard that can be maintained throughout the contract and be acceptable to the appointed supervisor. Only badly damaged bricks should be rejected such as would normally be discarded by the bricklayers.

Features such as soldier courses and narrow piers should be included so that matters related to brick size variation and workmanship can be resolved.

The reference panel described above should not be confused with a panel built to enable an architect to determine the type of mortar joint or other feature to be used or to establish achievable and acceptable standards of workmanship.

b) Sample panels

If it seems possible that the level of minor blemishes in a delivery may prove unacceptable, one or two pallets should be opened in the presence of the brick supplier's representative and the purchaser. A sample panel should be built in the same way as the reference panel where they can both be readily compared. If the brick supplier is not present when the packs which are the basis for complaint are selected and opened, it may be necessary to open more packs before an agreement can be reached. Arguments over the worst three bricks in the pack are seldom fruitful.

More information is given in *The BDA Guide to Successful Brickwork*, Section 1.1 'Building reference and sample panels'.

3.1.2 Avoiding colour banding

Colour banding or patchiness of brickwork may have three causes.

● The first is the result of variations in the general colour between different deliveries. If bricks have to be delivered over a long period, inform the supplier well before deliveries commence so that he can take action to minimize the risk of colour variations between one delivery and the next. This is important with both multi- and mono-coloured bricks.

● The possibility of colour banding or patchiness arising from variations within a single delivery should be dealt with as the packs are loaded out for the bricklayers as described in Section 3.2.5.

● If space permits, stock several loads of bricks on site to improve the chances of good colour blending. Site stocks should not be allowed to run out.

More information is given in *The BDA Guide to Successful Brickwork*, Section 3.2 'Blending facing bricks on site'.

3.1.3 Protecting facing bricks in the stacks

● Bricks should be stacked on firm, level bases and not in contact with soil, sulphate-bearing clinker or ashes and not where they are liable to be stained by mud and oil, splashed by vehicles or mortar mixing.

● Stacks should also be protected from rain as saturated clay bricks may contribute to efflorescence, lime leaching or both. See Sections 4.3.1 and 4.3.2. Saturated calcium silicate and concrete bricks will shrink more than dry bricks as they dry out and may cause cracking.

● Shrink wrapping of bricks by the manufacturer gives excellent protection, but provision should be made for protection of any deliveries not so protected

Fig. 3.2 Brick packs stacked off ground and protected by shrink wrapping.

and of packs once the wrapping has been removed or damaged (**Fig. 3.2**).

More information is given in *The BDA Guide to Successful Brickwork*, Section 1.3 'Handling, storage and protection of materials'.

3.1.4 Providing mortar of a consistent colour for the bricklayers

The importance of the colour of mortar to the final appearance of the brickwork is stressed in Section 2.2. It emphasizes the need to ensure that the colour of the mortar supplied to the bricklayers is consistent.

Mortar colour depends on the types of sand, cement, lime and pigments and the proportions in which they are mixed, as well as care in handling, storing and protecting the materials from contamination. The actions necessary to avoid variations in colour depend on which of the three basic types of mortar listed below are used.

a) Cement, lime:sand gauged and batched on site (Fig. 3.3)

Sand and cement should be obtained from consistent sources. The bricklayers' sand must not become contaminated by other sands, mud or oil. Store sand on a prepared, hard, clean base from which water can drain. Protect it from rain.

The proportions of cement, lime and

Fig. 3.3 Gauge boxes aid consistently accurate site batching of mortar.

sand should be accurately gauged and consistently batched. The consistence should be adjusted to suit the type of brick to be laid.

The addition of pigments on site is inadvisable, as accuracy when batching by volume is unreliable and will almost certainly lead to variations in mortar colour. If pigmented mortars are required it is safer to use pre-mixed lime: sand or ready-to-use retarded mortars as the addition of pigments can be more accurately controlled in a factory.

b) Pre-mixed lime:sand for mortar (coarse stuff) (Fig. 3.4)

Pre-mixed lime:sand should be accurately gauged with cement obtained from a consistent source. Adjust the consistency to suit the type of brick to be laid.

The mix should be carefully stored and protected. Particles of lime and pigments are very fine and can be washed out by rainwater or blown away from the edges in hot, dry weather.

(c) Ready-to-use retarded mortar (Fig. 3.5)

Containers in which ready-to-use retarded mortars are to be delivered should

Fig. 3.4 Ready-mixed lime:sand protected from the weather and contamination.

Fig. 3.5 Ready-to-use retarded mortar being discharged into containers.

be kept clean, covered and free from contamination. Inform the suppliers of the type of brick to be used so that the consistence may be adjusted correctly before delivery.

More information is given in *The BDA Guide to Successful Brickwork*, Section 4.1 'Mortars'.

3.2 Aspects of bricklaying that affect appearance

When the bricks and mortar have been specified, the detailed design completed, the site prepared and the construction programme planned it is down to the bricklaying team members, applying their knowledge and craft skills with care and attention to turn the concept into reality.

The BDA Guide to Successful Brickwork and *BDA Building Note 1*, 'Brickwork Good Site Practice', identify the many bricklaying operations necessary to ensure the successful performance of the brickwork. This section summarizes from both publications the salient points that affect appearance.

After a working lifetime concerned with the relationship between design and construction the author is convinced that a close liaison between designers, construction managers and craftsmen is a major factor in producing high quality work efficiently.

3.2.1 Setting-out facing brickwork and the 'rules' of bonding

a) Setting-out 'dry'

Before bricklaying begins, the brickwork should be carefully 'set-out' 'dry' at ground level in order to work out bonding in advance, particularly where the bond pattern will be interrupted by window openings and other features above. This is the time to identify problems and resolve them by discussion between the designers and the bricklayers.

b) 'Rules' of bonding

The following so-called 'rules' are no more than generally accepted good practice and are invariably followed unless other instructions are given.

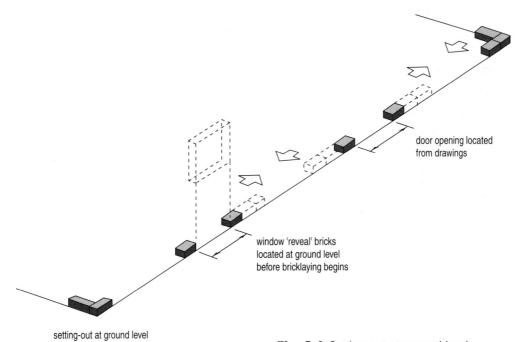

setting-out at ground level

door opening located
from drawings

window 'reveal' bricks
located at ground level
before bricklaying begins

Fig. 3.6 Setting-out at ground level.

Setting-out the brickwork 'dry' at ground level establishes first the positions of the 'quoin' bricks and then the 'reveal bricks' to all openings in the brickwork above (**Fig. 3.6**).

Running-out facing bricks 'dry' between 'quoin bricks' and between 'reveal bricks' enables:

- the bond pattern to be set out including the position of any necessary 'broken bond'
- perpends to be established for the full height of the brickwork in order to

avoid 'wandering' perpends where window openings occur at higher levels.

Symmetrical bonding between 'quoin bricks' and between 'reveal bricks' is the ideal (**Fig. 3.7**).

Quarter-lap bonds are normally established and maintained by cut bricks, called closers, being positioned next to the quoin or stopped-end headers. See Section 2.3.2 a.

Broken bond should be introduced if a length of brickwork is such that the bond pattern cannot be maintained

Fig. 3.7 Ideal dimensions – whole numbers of bricks and symmetrical reveals.

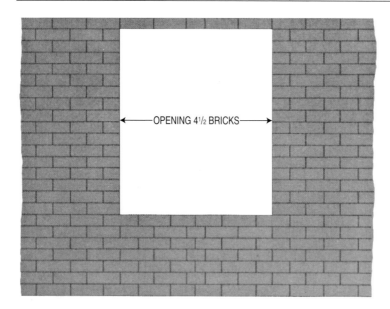

OPENING 4½ BRICKS

Fig. 3.8 Broken bond under centre line of window opening.

without unacceptably narrow or wide perpend joints.

Broken bond is normally positioned at the centre of walls and window openings. Alternatively broken bond at both ends may be preferred (**Fig. 3.8**).

Thin slices of brick, called bats, are generally unacceptable. In general, bricklayers work to the following 'rules'.

● No cut bricks less than a half-bat should appear on the face of the wall.

Closers, i.e. quarter-bats, should not be built in except after a header positioned at a quoin, stop end or reveal.

● If after setting out the brickwork, a quarter brick space is left, a three-quarter bat and a header should generally be used or two equally cut five-eighth bats if in Stretcher Bond.

Reverse bond is sometimes used to avoid broken bond (**Fig. 3.9**). A bond is said to be reversed when the bricks at

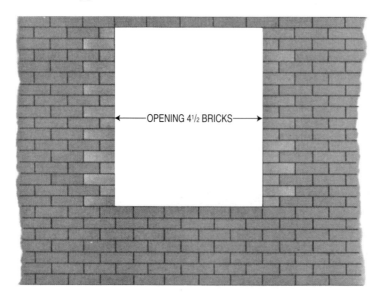

OPENING 4½ BRICKS

Fig. 3.9 Reverse bond avoids broken bond.

each end of a length of wall are not symmetrical. Bricklayers usually ask before using reverse bond as architects may hold strong if differing views on the matter. Reverse bond is likely to be more acceptable if the bricks and the mortar are of a similar colour, but less acceptable where bricks of a contrasting colour are used as dressings at window reveals.

'Perpends', the succession of vertical cross-joints, usually in alternate courses, should ideally rise vertically one above the other for the full height of the brickwork. But as the lengths of bricks vary it is not possible to do this through all vertical joints, neither is it necessary (**Fig. 3.10**).

In general, it is sufficient to plumb cross-joints at regular intervals, say every third or fifth perpend in stretcher half-bond. This degree of regularity is sufficient to avoid 'wandering' perpends which are visually so distressing.

Continuous vertical joints within the wall, known to bricklayers simply as 'straight jointing', should be avoided except where these are in the nature of the bond and inevitable, as in a one-brick wall in Flemish Bond (**Fig. 2.28**). The only time a straight joint should appear on facework is in Stack Bond. (**Fig. 2.47**).

At junctions and quoins tie bricks should securely connect the walls avoiding straight vertical joints between unless alternative methods of tying are approved.

3.2.2 Cutting bricks

Cuts through the face of a brick should be accurate and neat. 'Rough cutting' with a trowel is unlikely to be satisfactory. 'Fair cutting' with a club hammer and bolster can produce neat cuts in evenly fired bricks whereas bricks with fire cracks and other blemishes tend to shatter.

Multi-perforated bricks and very high strength, low absorption bricks can be very difficult to cut neatly by hand. For such bricks or a large number of bricks it is better to use a masonry bench saw. But see *The BDA Guide to Successful Brickwork*, Section 2.5 'Cutting Bricks' for more information including safety measures.

When cutting half-bats for use in an outer half-brick leaf of a cavity wall it might be preferable to expose the half stretcher face rather than a header face. The former would keep the appearance more consistent and avoid a rough cut projecting into the cavity.

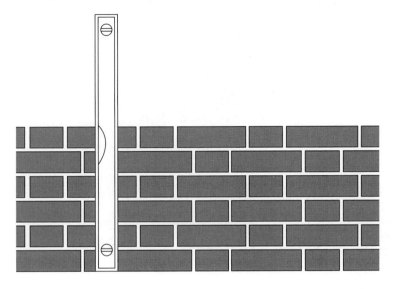

Fig. 3.10 Plumbing perpends.

3.2.3 Building to line, level, plumb and gauge

The appearance of brickwork is very dependent on the brick courses being 'to line' (straight), level and 'plumb' (vertical) and 'keeping to gauge'. The latter refers to building vertically between two given levels so that all courses rise the same amount above the one below and the bed joints are the same thickness. Modern standard bricks are built to a gauge of four courses to 300 mm. These are basic craft skills which every bricklayer should possess after good training and practice and conscientiously apply with care and attention.

After 'setting-out' normal straight brickwork as described in a) above, a bricklayer will raise quoins as control points, racking back to avoid toothing – it is very difficult to achieve a good appearance let alone solid, strong and rain resistant joints when building on to toothed brickwork.

3.2.4 Making the most of mortar joints

The surprising extent to which mortar joints affect the appearance of brickwork and the role of site management in providing mortar of a consistent colour is emphasized in Sections 2.2. and 3.1.4 respectively.

The bricklayer's role in making the most of this vital contribution to the appearance of brickwork is described below.

Newly built brickwork must be protected from rain to prevent fine particles of lime, cement or pigment being washed out, and changing the colour of the mortar joint (**Fig. 3.11**).

'**Jointing-up**' After each brick is bedded the surplus mortar or 'squeeze' should be cut off immediately, but the timing of the final finishing of the joints has to be carefully judged. If done too soon the wet mortar may be smeared over the face of the bricks. If left too late

Fig. 3.11 Protecting newly built brickwork from rain.

the mortar may crumble and lead to over-ironing of the joint which tends to blacken it. Mortar dries more slowly between low absorption bricks than between high absorption bricks, and more slowly in cold damp weather than in hot dry weather. The brickwork should not be brushed too early.

To maintain a uniform appearance bricklayers should use the same jointing techniques. Right- and left-handed bricklayers should strike back vertical cross-joints on the same side, usually the left hand side (**Fig. 3.12**).

Fig. 3.12 Striking back a vertical cross-joint.

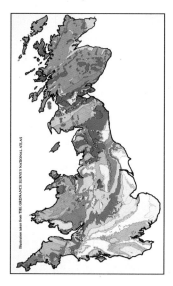

Plate 1 A variety of bricks stem from a complex geology. Reproduced from Ordnance Survey mapping with the permission of The Controller of HMSO © Crown copyright 87048M09/96.

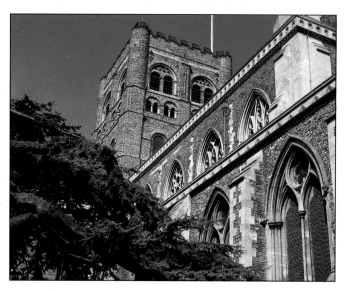

Plate 2 A Norman reuse of Roman bricks in the tower of St. Albans Abbey.

Plate 3 Hampton Court Palace 1515.

Plate 4 Stone dressing to a window at Hampton Court Palace.

Plate 5 Elaborately moulded chimneys at Hampton Court Palace.

Plate 6 Kew Palace, formerly known as the 'Dutch House' 1631– one of the first uses of Flemish Bond.

Plate 7 House in Bedford Square, London W1, c.1780 – a prime example of the regular, disciplined town housing common from the early 17th to the early 19th centuries.

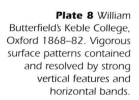

Plate 8 William Butterfield's Keble College, Oxford 1868–82. Vigorous surface patterns contained and resolved by strong vertical features and horizontal bands.

Plate 9 Office training centre on a mature site, built of dark, irregular multi-rough stocks to 78 mm rather than a 75 mm gauge. Irregularities were permitted in front of the bricklayer's line.

Plate 10 By contrast with plate 9, the same architect used light-coloured, precise, calcium silicate bricks to complement the open steel framing at a Sixth Form Centre on an open site.

Plate 11 A textured wall of 'common' Fletton bricks laid fairfaced displays boldly coloured tapestries.

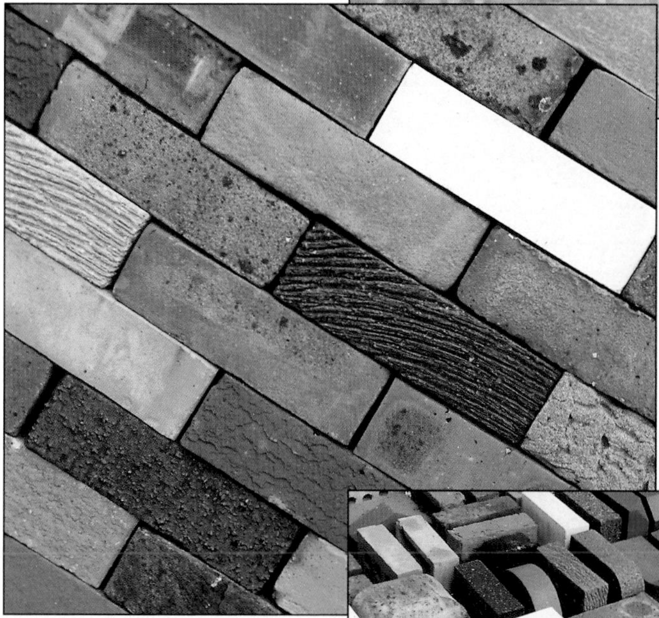

Plate 12 Examples from the range of brick colours and textures available.

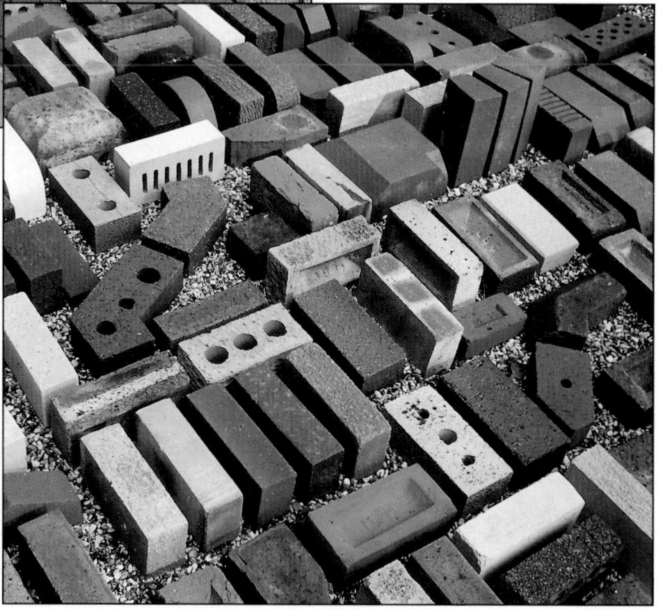

Plate 13 A variety of brick forms and shapes.

Plate 14 Low-relief carvings in the County Hall, Morpeth, Northumberland.

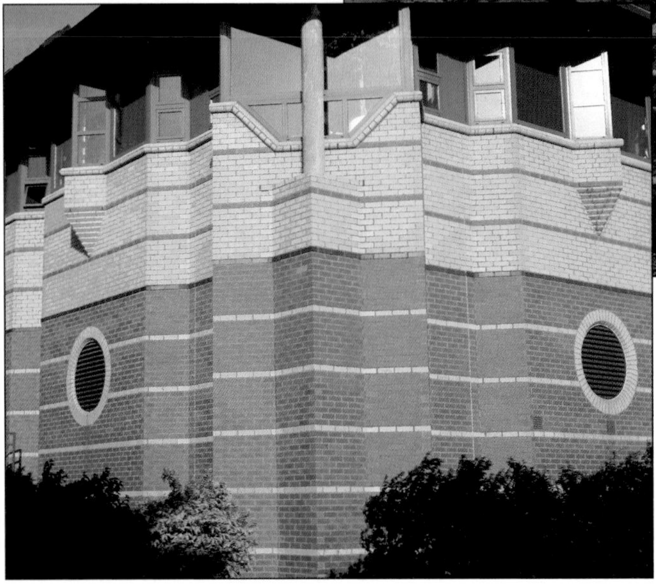

Plate 15 Polychrome brickwork, West Yorkshire Playhouse, Leeds.

Plate 16 Functional buttresses to wall of machine hall and mechanical laboratories make delightful architectural features with restrained red bands.

Plate 17 Glazed blue bricks, Westbourne Grove, London W2.

Plate 18 Magistrates Court Cardiff. Detail showing precise Stack Bonding of piers with purpose-made 'specials' at the angles and soldier courses. To meet the need for bricks having fine tolerances the brickmaker gauged 20 000 bricks before delivery.

Plate 19 'The Circle' – distinctive use of glazed blue bricks in housing, London SE1.

Plate 20 A bold use of glazed blue bricks in contrast with white adjoining buildings – Atlantic House, Wardour Street, London.

Plate 21 Glazed blue bricks set jewel-like in a wall of red, textured bricks in the Croydon library complex.

Plate 22 Glyndebourne
Opera House, East Sussex.

Plate 23 Detail of structural
brickwork, Glyndebourne
Opera House.

Plate 24 Interior brickwork,
Glyndebourne Opera House.

Plate 25 Coloured pointing mortars form patterns in brickwork at the Quinton and Kynaston School, London NW8.

Plate 27 The Granary, Welsh Beck, Bristol, 1869. English Garden-Wall Bond and polychrome brickwork.

Plate 26 Irregular Tudor brickwork with wide joints next to smooth red bricks with fine joints in the brickwork of Christopher Wren's Hampton Court Palace.

Plate 28 Yellow stock bricks in Flemish Bond with diaper patterns in red bricks. Old Pye Street, London SW1.

Plate 29 Curved wall of auditorium and vertical ventilation stacks in yellow calcium silicate bricks with blue headers. Engineering School, De Montfort University, Leicester.

Plate 30 Town Hall, Reading, Berkshire 1872–75.

Plate 31 Rugged textured, hand-made bricks with band courses of stone at St Brigid's Church, Belfast.

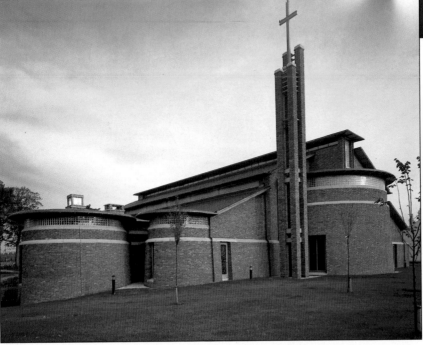

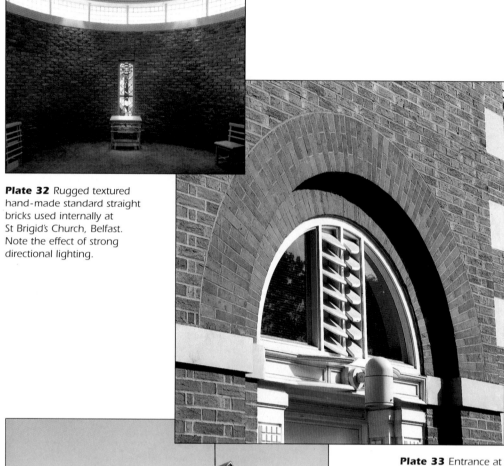

Plate 32 Rugged textured hand-made standard straight bricks used internally at St Brigid's Church, Belfast. Note the effect of strong directional lighting.

Plate 33 Entrance at Mercer's House showing two-ring semi-circular arch with fine joints between pre-formed voussoirs and 50 mm high bricks in Flemish Bond using snapped headers.

Plate 34 A bold use of multi-coloured bricks including plinth 'specials' at the River Foss flood alleviation scheme, York.

Plate 35 Simple massing with unassertive detailing lends a quiet dignity to the school dining hall at Felsted School, Essex.

Plate 36 An elegantly modelled stack at Queen Mary's Hospital, Hampstead.

Plate 37 A niche in gauged brickwork – restored.

Plate 38 Bricks ranging from brown through orange to pink set off by stone features and dressings at Mercer's House, London N1.

Plate 39 Yellow brickwork and stone banding harmonise with adjacent buildings at Richmond House, Whitehall, London.

Plate 40 A series of retaining walls formally delineate the fall from the footpath to the buildings below.

Plate 41 Flexible paving of 200x100x65mm clay pavers in just two colours at Wakefield.

Plate 42 The new paving relates and unifies Wakefield city and the cathedral.

Plate 43 Mainly strong red pavers with some blue and buff in 45° basket weave pattern at Huchnall and Newstead stations, Robin Hood line, Notts.

Plate 44 Clay pavers in Herringbone Bonded flexible paving subject to heavy traffic in Oldham.

Plate 45 Brindle, blue and buff pavers laid out in Centenary Square, Birmingham, resemble a collection of oriental carpets.

Plate 46 At New Street in Huddersfield, the buff pavers mark pedestrian routes known to be preferred while the brown pavers form a meandering line to avoid reinforcing the linearity of the existing street.

Plate 47 A carved brick
panel from 'The Creation'
at the Eye Hospital, Bristol.

Plate 48 Carved
inscription on brick
bridge in Taunton.

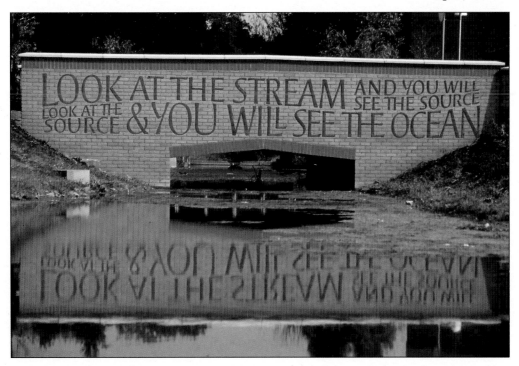

Plate 49 House near Exeter with simple bands and cappings of brick-on-edge and brick-on-end. The splayed corners are generated by 'special' angle bricks at 45° angles.

Plate 50 Bold curved walls appropriately complemented by robust and stone copings and string courses at Spedan Tower, Hampstead.

Plate 51 High-rise flats, Northwood Tower, Walthamstow. Refaced with bricks.

Plate 52 Thames Water pumping station, Isle of Dogs, London.

Plate 53 Offices required by the planners to be in the form of the 1787 church demolished as structurally unsound. New structural brickwork, in Flemish Bond. Grimaldi Park House, Pentonville Road, London.

Plate 54 Cladding for PowerGen's Coventry offices. Extruded wire-cut bricks were Stack Bonded to align perforations to accommodate vertical prestressing tendons (cables).

Plate 55 Alternating two-course bands of red and buff bricks with reconstructed stone features give a striking appearance to 200 Aztec West, Bristol.

Plate 56 A structural fin wall encloses and enhances the Leicester University Sports Hall.

Plate 57 Student accommodation in the Arco building at Keble College, Oxford.

Plate 58 Detail of multiple soldier courses in the Arco building.

Plate 59 The effect of the boldly contrasting bands of brick is intensified by matching the mortar to the brick in each band at The Katherine Stephen room, Newnham College, Cambridge.

Plate 60 St Mary's Cathedral, Liverpool.

Plate 61 St Lazars Hall for the Serbian community, in Bournville, Birmingham. Bands of dark red bricks in 20 mm thick mortar joints, made with white sand. Brickwork alternates with bands of blockwork in lieu of the traditional natural stone.

Plate 62 St Lazars Hall. Further embellishment includes courses of dogtoothing and inclined soldier courses on the tower. Arch voussoirs contain tapered joints.

Plate 63 Refurbishment of Bury St Edmunds Railway Station.

Plate 64 Refurbishment of Bury St Edmunds Railway Station, detail.

Plate 65 Multi-stock bricks separated by single courses of blue bricks from the stone bands at the Clayton Square shopping centre in Liverpool.

Plate 66 Domestic chimney stack with decorative quoins and a stone coping.

Plate 67 Restrained modelling with plinth bricks, corbelling and surface patterning helps to humanize sheltered housing in Liverpool.

Plate 68 'The Steelworker' – a 13 m high polychromatic mural in Sheffield, using 18 types of brick and five coloured mortars.

Plate 69 Segmental, unreinforced brick bridge spanning 8 m replaced a steel joisted structure in 1992 at Kimbolton, Cambridgeshire.

Plate 70 Reinforced brickwork retaining walls at the Bottiler subway at Banbury.

Plate 71 Post-tensioned brickwork diaphragm abutment walls for the Foxcovert railway bridge, Glinton by-pass, Cambridgeshire.

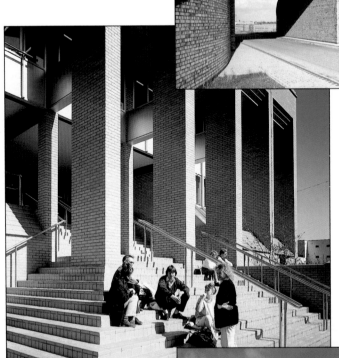

Plate 73 Three-ring arch – Oceanography Centre, Southampton.

Plate 72 Royal Scottish Academy of Music and Drama, Glasgow.

Pointing as described in Section 2.2.4, requires considerable knowledge, skill, experience, care, attention and patience. It is best done by specialists. The appearance of attractive old brickwork can be ruined by insensitive, unskilled repointing and old bricks can deteriorate as the result of inappropriate repointing. For further details see *The BDA Guide to Successful Brickwork*, Section 2.8 'Pointing and repointing'.

3.2.5 Avoiding colour banding and patchiness

Within any delivery of bricks, whether they are multi-coloured or mono-coloured, there may be some variants differing slightly in colour from the majority. They may be dispersed within individual packs or be concentrated in particular packs, or both.

To prevent the variants becoming concentrated in patches or bands rather than being dispersed throughout the brickwork, the bricks should be 'loaded out' to the bricklayers from four or five packs at a time (**Fig. 3.13**). Preferably, the bricks should be taken in vertical rather than horizontal slices as the position of bricks in a pack tends to match their position vertically in the kiln. Some manufacturers recommend that the bricks be drawn in a diagonal line across a pack.

Where sites are too confined to carry out blending on site many manufacturers are able to blend in the factory. If this service is required it should be discussed with the suppliers at an early stage.

3.2.6 Selecting bricks for uniformity of size

Features such as soldier courses, arches, copings to one-brick thick walls and narrow piers require bricks with more uniformity of size than can be expected of bricks supplied to the limits of size specified in BS 3921: 1985. Bricks for such features must be selected by gauging, to a variation about the mean size of no more than ± 2 mm. Selection can be made either on site or by the supplier before delivery.

More information is given in *The BDA Guide to Successful Brickwork*, Section 6.5 'Allowing for variations in brick sizes'.

3.2.7 Bricklaying and the weather

Anticipation and simple precautions can avoid the consequences of extremes of weather conditions on the appearance of completed brickwork.

a) Bricklaying in the rain

Newly built brickwork should be protected from even light rainfall and run-off to avoid 'lime leaching' and 'efflorescence', see Section 4.3. Protection from saturation should be maintained after the mortar has hardened until completion. For more information see *The BDA Guide to Successful Brickwork*, Section 1.2 'Protection of newly built brickwork'.

b) Bricklaying in cold weather

Bricklaying should cease when a falling air temperature reaches 3°C.

Note: If mortar freezes before it sets, its ultimate adhesion, strength and long term durability will be reduced and the brickwork will probably have to be rebuilt when care must be taken to ensure that it matches.

'Anti-freeze' admixtures are of little value for masonry mortars and are not recommended. For more information see *The BDA Guide to Successful Brickwork*, Section 3.1 'Avoiding damage from extremes of weather'.

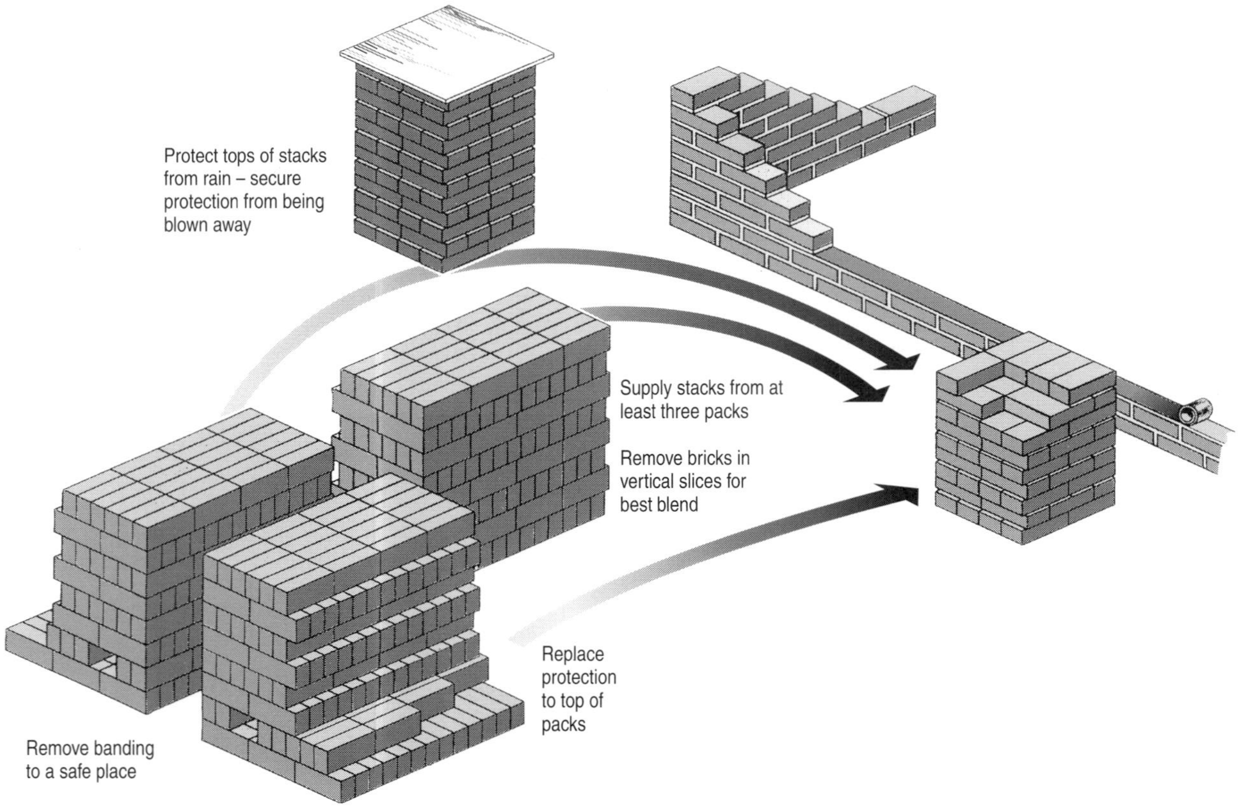

Protect tops of stacks from rain – secure protection from being blown away

Supply stacks from at least three packs

Remove bricks in vertical slices for best blend

Replace protection to top of packs

Remove banding to a safe place

Fig. 3.13 'Loading out' to bricklayers from at least three packs.

c) Bricklaying in warm drying weather

Newly built brickwork should be protected from the combined effects of high temperature, wind and highly absorbent bricks drying out the mortar before it has set and properly bonded with the bricks. Deteriorating mortar joints will spoil the appearance as well as weaken the brickwork. For more information see *The BDA Guide to Successful Brickwork*, Section 3.1 'Avoiding damage from extremes of weather'.

3.2.8 Keeping brickwork clean and undamaged

Clean facework, free of mortar smudges separates the skilled and conscientious bricklayers from the rest and begins with good trowel technique and mortar of the right consistence. Excessively wet mortar is a common cause of dirty brickwork.

- A skilled bricklayer will spread just the right amount of mortar evenly along a course and bed the bricks with gentle pressure. Too much tapping encourages water towards the face, down which it dribbles.
- The mortar joints should be finished with care. See Section 3.2.4.
- The scaffold boards closest to the wall should be turned back when work stops to prevent rain splashing the brickwork with mortar. Plinths and other projections should be protected from mortar droppings from bricklaying above and other vulnerable brickwork, such as reveals to openings, should be protected from physical impact.

Working cleanly has been covered in detail in *The BDA Guide to Successful Brickwork*, Section 2.6 'Keeping brickwork clean'.

3.2.9 Building with special shapes

Bricks of special shapes and dimensions and their uses are described in Section 2.1.1 g. Craft matters concerning bricklaying craft are dealt with in *The BDA Guide to Succesful Brickwork*, Section 2.9 'Building with bricks of special shapes and sizes'. Three particularly important points are set out below.

- 'Specials' should be carefully looked after as, besides being more expensive than standard bricks, they are made to order, and cannot always be replaced from stock.
- Left- and right-hand versions are essential for some special shapes.
- But although the shape of some specials apparently allows them to be used for both left- and right-handed situations by simply inverting them there may be good reasons why this should not be done. See Section 2.1.1 g (ii) and **Fig. 2.6**.

3.2.10 Building vertical movement joints

The need for movement joints is explained in Section 2.4.3 f. This section highlights some aspects of particular concern to bricklayers.

- The bed joints to the faces within the movement joints should be finished flush so that the joint can be sealed effectively and neatly.
- Maintain verticality and a constant joint width. Movement joints are usually very conspicuous.
- Mortar droppings and debris should be removed from the joint before placing the filler and sealant.
- Filling and sealing movement joints is best left to specialists but on small jobs it is often left to the bricklayer.

3.2.11 Building banded brickwork

The eye is readily attracted to any soldier courses and so extra care must be taken to set the bricks plumb. An out-of-plumb soldier brick is as conspicuous as a guardsman in the front row fainting on parade. Generally, every third soldier brick is plumbed with a boat level.

Alternative ways of returning a soldier course at a corner are shown in **Fig. 2.56** in Section 2.4.2.

4 Weathering of brickwork

4.1 Causes

With the passage of time the appearance of brickwork changes as the result of either a fairly even, patina-like deposit of detritus from the atmosphere or the distribution of the detritus and other matter over various surfaces by run-off rainwater.

4.1.1 Patina-like deposits

This generally results in a 'softening' of colours and is often judged to enhance the appearance of brickwork, although in polluted atmospheres deposits may in time obscure the colours and textures. Then expert cleaning will be necessary to refresh the appearance.

4.1.2 Rainfall run-off

The rate at which rainfall runs off a surface depends on how absorbent it is. Virtually all the rainwater falling on windows and impervious claddings as well as very low absorption bricks will flow as soon as rain falls, the 'raincoat' effect. Whereas, if the wall is of absorbent bricks much of the initial rain will be absorbed to dry out later, the 'overcoat' effect.

It is advisable to protect brickwork from run-off by an overhanging feature such as a sill with a throat or other form of drip underneath. If the amount of run-off is very large a gutter is preferable.

Weathering caused by rainwater run-off often leads to uneven streaking and is probably more disfiguring than an even patina. Contaminated water flowing over a surface may deposit solid particles of foreign matter, but otherwise may keep a surface clean compared with adjacent less well-washed surfaces.

Concentrated run-off may therefore result in 'dirty' or 'clean' streaks. They may be caused by a design feature such as mullions in a window, ends of window sills that are not stooled or by faulty joints in copings or sills that allow water to by-pass the throat and run down in concentrated streams. Water running down a vertical surface often tracks sideways under the influence of wind rather than running down vertically. Streaking is often more noticeable on smooth surfaces which do not disperse a stream of run-off as does brickwork which tends to be more textured and absorbent.

Flowing water may dissolve soluble salts from some materials and then deposit them on the brickwork leaving unsightly stains. Examples are the white carbonate which forms on the surface of new lead sheet but stains brickwork below black or yellow depending on the conditions. It can be prevented by the application of patination oil to the lead as soon as it is fixed. Washings from copper are noticeable as a green stain, particularly on light coloured bricks. Salt deposits from cementitious materials may also cause stains.

The extent of weathering will also depend on the degree of exposure to wind-driven rain which in turn depends on geographic location and the position of nearby buildings and terrain. The most exposed brickwork will tend to be on relatively high rather than low terrain; facing the prevailing winds; in the west of Britain rather than the east; high above the ground and facing open spaces rather than nearby protective buildings and trees.

For a particular building the quantity of rain initially falling directly on its various 'collecting' surfaces depends on the intensity and duration of the rain and the wind direction and speed.

4.2 Some examples of weathering

Figs. 4.1–4.9 illustrate some typical examples observed in practice.

Fig. 4.1 There is less run-off over the concrete beam from the relatively absorbent brickwork than from the glazing on either side.

Fig. 4.2 The gutterless glazed roof adds to the run-off down the vertical glazing and the mullions concentrate it over the inadequate sill, saturating the masonry in vertical strips resulting in algae growth.

Fig. 4.3 Mullions concentrate wind blown, sideways-tracking run-off from the glass over the brickwork beneath, helped by the lack of a throated sill.

Fig. 4.5 Concentrated run-off from a bull's-eye window causes efflorescence or lime leaching and could cause frost attack in M quality bricks.

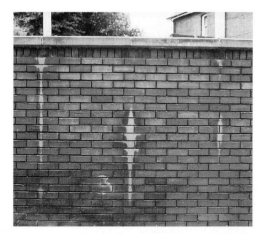

Fig. 4.4 Faulty coping joints allow run-off from coping to by-pass throating to underside of overhang and concentrate down the face of the wall.

Fig. 4.6 Another designer has devised a detail to allow the run-off to drip clear.

Fig. 4.7 The top of a freestanding wall with a flush brick-on-edge capping remains wet long after rainfall ceases in contrast with a wall protected by a projecting and throated coping.

Fig. 4.8 Stock type clay bricks in freestanding wall are darker than the same bricks in adjacent external wall of building. Each was built 16 years before being photographed.

Fig. 4.9 Brickwork of low absorption bricks and flush detailing after ten years.

4.3 Saturation and deterioration

Excessively saturated brickwork will not only be unsightly but may lead to physical deterioration and affect long term durability as described below. It may all be avoided by appropriate design and specification.

4.3.1 Efflorescence

This is a white fluffy deposit on the surface of clay brickwork as the result of saturation dissolving soluble salts from the brickwork which are deposited as the water evaporates.

Efflorescence normally appears during the first drying weather after construction and, although not pleasing to see, is generally quite harmless and being soluble usually disappears within a month or so unless the brickwork remains saturated.

The risk of efflorescence can be minimized by detailed design to prevent saturation, by protection of bricks stacked before construction, during construction and by good maintenance of rainwater goods, flashings and copings.

4.3.2 Lime leaching

Lime leaching is a hard white insoluble deposit resulting from free lime carbonating on the surface of brickwork. The deposit is difficult to remove except by chemical cleaning agents. Free lime may leach from newly built saturated brickwork or, in more mature brickwork, from adjacent saturated mortar, concrete or other cementitious materials.

The risk of lime leaching can be reduced, first by effective damp-proof membranes to isolate the brickwork from other materials such as external concrete floor and step structures, and second by protecting newly built brickwork from rain.

4.3.3 Lichen, algae and mosses

These can grow on external brickwork that has a fairly high and constant level of water, mineral salts for food, and light. More about these and other growths including remedial measures are given in *BRE Digest 370*, March 1992, 'Control of lichens, moulds and similar growths'.

The enthusiasm of some for the appearance of such growths, especially on old buildings, may not be shared by those responsible for their upkeep if they recognize it as a sign of long term dampness and deterioration.

4.3.4 Frost attack

'F' quality (frost resistant) clay and most calcium silicate and concrete bricks are not affected by frost attack. It can occur in 'M' quality (moderately frost resistant) clay bricks only if the brickwork is saturated when the temperature falls below the freezing point of water. Frost attack will not occur in 'M' bricks if the brickwork is protected from saturation by appropriate details.

4.3.5 Sulfate attack

Sulfate attack in mortar joints is rare and occurs only if there is a considerable amount of water movement through the brickwork. Diffusion alone is not sufficient. Soluble sulphates, usually from the clay, but possibly from other sources react with tricalcium aluminate (C_3A) in the cement. This causes an expansive crystaline growth within the mortar joints causing them to crumble, erode and, in severe cases, expand and crack. Earth-retaining walls are particularly vulnerable if not properly designed and constructed. See Section 2.6.2.

4.3.6 Preventive design

Run-off rain can be intercepted by sills, gutters and copings to prevent the saturation, disfigurement and deterioration of surfaces below.

Particularly vulnerable features include the tops of freestanding, retaining and parapet walls and chimney stacks. In particular a damp-proof membrane must be applied to the earth-retaining surface of retaining walls to prevent the saturation which can lead to disfigurement as well as physical deterioration from efflorescence, lime leaching and sulfate attack.

5 Brick manufacture

Knowing something of the way bricks are made helps us to understand and appreciate the variety of colours and textures. The following describes briefly the principal brickmaking raw materials, the way they are processed and how both affect appearance as well as other physical characteristics including strength, absorption and durability. A more comprehensive illustrated description is given in *The BDA Guide to Successful Brickwork*.

5.1 Clay bricks

5.1.1 Types of clay and types of bricks

a) Types of clay

The clays from which we make bricks were laid down in a variety of conditions from 400 to 1 million years ago. In time, a disintegrating, twisting, cracking and slipping earth's surface resulted in a complex geology and a profusion of different brickmaking clays. See Table 5.1.

Carbonaceous clays contain fuel in the form of carbon. The Lower Oxford clay, is a highly carbonaceous clay found deep under the superficial deposits of Oxford clay in Bedfordshire and Bucks. Fletton bricks which are made from this clay are named after the village of that name near Peterborough, then in Huntingdonshire, where 'the clay that burns itself' was first discovered. Coal measure shales are also carbonaceous and are found mainly in the Midlands and North but are more scattered than the Lower Oxford clay.

Non-carbonaceous clays such as Etruria Marl, Keuper Marl, Weald clay and Boulder clay are often red burning while others such as Gault and Lias clays which may have a high lime content might burn cream or white, particularly if further lime is added to the clay.

b) Some types of bricks

Red burning clays may be fired blue, brindled or brown by controlling the amount of air admitted to the kilns. For example Staffordshire Blue bricks are produced from Etruria Marl and fired with the atmosphere in the kiln controlled to achieve reducing conditions in which the oxygen content is limited.

Multi-coloured stock bricks were originally the result of adding decayed and combustible town refuse to some non-carbonaceous clays but generally fuel is now added in the form of crushed coke.

Yellow London Stock bricks are similarly made but by adding chalk as well as fuel. After firing they are sorted according to the hardness of firing into first hard stocks (just what the manufacturer was aiming for), second hard stocks (not

Table 5.1

	Age (millions of years)	Environment of deposition	Area used for brickmaking
Boulder clay	< 1	Glaciers & rivers	N England
Brickearth	< 1	Land	SE England
Reading Beds	circa 65	Deltas	S England
Gault clay	circa 90	Marine	SE & E Anglia
Weald clay	circa 130	Deltas/lakes	SE England
Oxford clay	circa 160	Marine	Central/E Anglia
Keuper Marl	circa 180	Salt lakes	Midlands
Etruria Marl	circa 270	Fresh water	Midlands
Coal measure shales	circa 310	Deltas & swamps	S Wales, Midlands, N Scotland
Coal measure Fire-clays	circa 310	As above	S Wales, Midlands, N Scotland
Culm measure	circa 310	Marine	SW England
Devonian Shale	circa 390	Marine	SW England

quite perfect), mild stocks (slightly under-fired and paler in colour), roughs (over-fired and somewhat misshapen) and commons (not intended for facing brickwork).

5.1.2 Winning and processing clays

Where different types of clay are to be mixed it is common to stockpile them in horizontal layers. In time the clay is removed by taking full vertical cuts from the face, in order to blend the clays to improve consistency and minimize the risk of colour variations. The clay is then transported to the brick plant and prepared for the shaping process.

5.1.3 Preparing the clay for shaping

To maintain consistency of colour and texture the clay is carefully prepared to avoid variations in content, grading, plasticity and water content. The fineness of grinding affects the textures of bricks. The largest particles are now typically 3.5 mm in diameter. Bricks having a smooth face are manufactured from clay that has been more finely ground. The effect of particle size is par-

ticularly evident in the appearance of dragface products.

5.1.4 Additions to the clay that affect the appearance

Materials may be added to modify the body colours. For example, a clay that would normally burn red will burn brown or grey if manganese dioxide is added. Sand or coke may be added to create a specific appearance, to control shrinkage or to provide fuel. If lime is added to 'soft-mud' clays to improve their workability the colour may be lightened.

5.1.5 Shaping clay bricks

The majority of clay bricks in the UK are made by one of three processes which have evolved to suit clays with fundamentally different moisture contents and workabilities.

a) Soft-mud method of shaping – by hand and machine

i) Hand-making (Fig. 5.1)

The crease patterns characteristic of hand-made bricks, sometimes known as 'the handmaker's walk', show a 'smile'

Fig. 5.1 Hand throwing in progress.

Fig. 5.2 A column of clay being extruded.

on the face when laid in the wall with the frog upwards. If the frog is laid downwards the down-turned ends of the crease appear to disapprove.

The insides of the moulds are sand coated to aid the release of the brick from the mould but the sand also imparts a colour and texture to the fired brick.

ii) Machine moulded

Typically a 'soft-mud' clay mix is forced through steel dies into sanded moulds. The faces of machine moulded bricks are smooth rather than creased like hand-made bricks.

A development of this process simultaneously throws sanded clots into a series of moulds. This simulation of the hand-throwing process produces bricks that have crease marks difficult to distinguish from hand-made bricks.

iii) Slop moulding

Bricks are made from a 'soft-mud' mix but are released from the moulds by water rather than sand, giving a different surface texture which, on firing, reflects the colour and texture of the clay rather than the sand.

b) Extruded and wire-cut method (Figs. 5.2 and 5.3)

In the UK about 40% of bricks are made this way. A plastic clay mix is driven through a die forming a continuous extruded column having a cross section

Fig. 5.3 Wire cutting a column of clay.

based on the length and width of a brick.

Surface textures can be introduced at this stage by various methods. These include brushing; removing the surface with a 'drag wire' which partially reveals some of the larger pieces of aggregate; rolling on patterns and coating with sand. Colours may be added by stains applied with the sand or via sprays. The column is then cut by wires into brick heights.

Extruded wire-cut bricks may be repressed to obtain sharper corners, a particular surface finish or chamfers etc.

c) Semi-dry pressed method

Facing bricks made by this method are either sand-faced or textured. The term 'semi-dry' refers to the workability of the clay at its naturally low moisture content. A fine granular material is produced by grinding the raw clay which is then pressed into bricks. Some 20% of UK bricks are made this way, mainly from the Lower Oxford clay but also from some colliery shales.

5.1.6 Drying clay bricks

Most clay bricks have to be dried to reduce the moisture content before firing and it is at this stage that the green brick may shrink by 5–14% depending on the type of clay and the moisture content. This process is carefully controlled to prevent cracking, bowing and twisting. Clay bricks after drying, but before firing, are referred to as being in their 'green' state.

5.1.7 Firing clay bricks in kilns

Strength, hardness, durability and colour are imparted to the moulded and dried green bricks by firing them at high temperatures in kilns which today generally burn gas, whereas in the past, oil, coal, coke and even wood have been used. The green bricks undergo a sequence of fundamental chemical and mineralogical changes as the temperature rises during the firing process.

Between 200 and 900°C, carbonaceous material which may be naturally present in the clay or added for fuel, burns out and can be an important contribution to the colour. Between 900 and 1250°C, liquids form as some components of the clay minerals melt. On cooling, they form a glass which binds the clay particles to form a hard and durable brick.

At this stage the characteristic colours develop. Normally the majority of clays fire red but fire-clays for example will yield buff/cream coloured bricks. These basic colours can be modified by adjusting the fuel/air ratio and by ensuring that some carbon is retained in the body up to the highest temperatures. Further shrinkage takes place during firing of up to 7% in addition to that which takes place during drying.

Kilns can be broadly categorized into two types, 'continuous' and 'intermittent'.

a) Continuous kilns

There are two types of continuous kilns.

i) Tunnel kilns (Fig. 5.4)

More than 60% of all UK bricks are fired in tunnel kilns which consist of a long narrow tunnel through which rail-mounted 'cars', loaded with pre-dried-'green' bricks are pushed at regular intervals. Typically, they take 2½ days to pass through, during which peak temperatures and rates of heating and cooling can be adjusted to suit a particular type of brick. Tunnel kilns may be controlled manually but are today more often electronically controlled and may be managed by computer programmes.

ii) Chamber kilns

A chamber kiln consists of a continuous circuit of inter-connecting chambers in which bricks are both dried and fired. The fire is continually drawn round the circuit from one chamber to the next and when the fire has moved on and the bricks have cooled they are unloaded. The larger kilns have two fires circulating at a time. About 25% to 30% of bricks are fired in such kilns, the vast majority being Fletton bricks made from the Lower Oxford clay.

Fig. 5.4 Cars of fired bricks about to emerge from a tunnel kiln.

b) Intermittent kilns

About 8% of clay bricks are still fired in intermittent kilns where appropriate for the scale of the operation and the type of brick being made. Intermittent kilns consist of a single chamber into which bricks are loaded for firing and withdrawn once they have cooled. Control is largely in the hands of experienced and skilled kiln burners.

A few manufacturers still use **clamps** which consist of large stacks of green bricks traditionally set on a bed of coke. The green bricks contain sufficient added fuel in the form of coke and combustible refuse to raise the temperature to about 1100°C when fired. Combustible town refuse was used in the past but since it no longer contained sufficient combustible material crushed coke has been used. Gas firing is also used today for this process.

Once a clamp is ignited little can be done to control it and maintain consistent firing conditions throughout is length, width and height. Consequently some parts reach higher temperatures than

others producing a wide range of colours and textures.

Once cooled, after burning for a number of weeks, the clamps are dismantled and the bricks visually sorted. Those round the edges of the clamp may be pink and soft through being underburnt. These have always been known as 'sammel' bricks and were not suitable for exposure to the weather. The remainder being anything from adequately and properly burnt to overburnt, vary greatly in colour and texture, a characteristic which is eminently suitable for some buildings and localities (**Fig. 5.5**).

5.2 Calcium silicate bricks – composition and manufacture (Fig. 5.6)

Calcium silicate bricks, having been first patented in the UK in 1866 by an American, were first produced in Germany in 1894 and in the UK in 1904.

Calcium silicate bricks are also known as 'sandlime' or 'flintlime' bricks. Both are made from a mixture of sand and hydrated lime but with some crushed flint added for flintlime bricks. Iron oxide pigments are added as required. The aggregates account for about 90% of the dry weight of the mixture.

Fig. 5.5 Firemarks on clamp-fired bricks.

Fig. 5.6 Calcium silicate bricks in an autoclave.

Enough water is added to allow the mixture to be moulded under high pressure.

The 'green' bricks are then subjected to high-pressure steam in an autoclave in which the lime reacts with the silica to form hydrated calcium silicates producing durable strongly bonded bricks. The process today is highly mechanized and subject to rigid quality control at all stages. The physical characteristics of calcium silicate bricks are covered in Section 2.1.2.

5.3 Concrete bricks – composition and manufacture

Concrete bricks are composed of aggregates such as crushed granite and limestone bonded with cement and coloured with iron oxide pigments. Generally the semi-dry concrete mix is compacted in mechanical vibratory or hydraulic presses. The water content is strictly controlled to ensure bricks with a consistent strength and water absorption. Like calcium silicate bricks they have a crisp precise appearance. The physical characteristics of concrete bricks are covered in Section 2.1.3.

Glossary

*Terms printed in *italics* in the definitions are separately defined within this glossary.

Actual size – the size of an individual brick or block as measured. It may vary from the *work size* within certain allowances for *tolerance* (see also *coordinating size* and *work size*).

Alluvial clay – deposits left by flowing water, e.g. in river valleys and deltas.

Angle bricks – *special* shape bricks which form non right angled corners in walls.

Arch – an assembly of bricks which spans an opening in a wall. Usually curved in form, but may be practically flat.

Arch bricks – *special* bricks for use as *voussoirs* in semi-circular *arches* of designated spans and in appropriate segmental arches.

Arris – any straight edge of a brick formed by the junction of two faces.

Ashlar – *masonry* consisting of blocks of stone, finely square-dressed to given dimensions and laid in *courses* with thin joints.

Autoclave – a pressure vessel used in the manufacture of *calcium silicate bricks* in which they are subjected to super heated steam at high pressure.

Axed arch – an *arch* formed of bricks cut to appropriate wedge shape by the bricklayer (see also *gauged arch*).

Band course – one or more *courses* forming a decorative contrast of colour, *bonding* or shape. The courses may be flush, projecting or recessed.

Bat – a part brick, e.g. half-brick, three-quarter brick, used in *bonding* brickwork at corners and ends of walls and to break bond. See *broken bond*.

Batching – the accurate proportioning of *mortar* materials to produce a specified mortar mix.

Battlements – *parapet wall* with raised sections to protect the defenders.

Bed – the layer of *mortar* on which a brick is laid and supported.

Bed face – the face(s) of a brick usually laid in contact with a mortar *bed*.

Bed joint – a horizontal joint in brickwork.

Bench saw – a power-driven, circular saw mounted on a bench – in the context of this book for cutting bricks.

Bolster – a broad-bladed chisel of hardened steel used for cutting bricks.

Bond(ing) (1) – the arrangement of bricks in brickwork, to maximize strength and robustness and generally to show a regular pattern in *facework*.

Bond (2) – the resistance to displacement of individual bricks in a wall provided by the adhesive function of *mortar*.

Bonding bricks – part bricks, e.g. half or three-quarter bricks, or specially shaped units to facilitate *bonding* of brickwork at features, corners and ends of walls (see also *bat*).

Boulder clay – a type of clay formed by glacial action. It contains mixed sizes of particles from fine clays to boulders.

Brick – see *calcium silicate (sandlime, flintlime)*, *clay*, *common*, *concrete*, *engineering*, *extruded wire-cut*, *facing*, *Fletton*, *handmade*, *perforated*, *pressed*, *semi-dry pressed*, *soft-mud (handmade)*, *stock* bricks.

Brickearth – silty clay or loam in a shallow deposit. Traditionally used for making clay bricks.

Brick slips – see *Brick face slip* and *Brick bed slip*.

Brick face slip – thin slice of brick to simulate a *stretcher* face.

Brick bed slip – thin slice of brick to simulate a *bed* face.

British Standards – national standards defining the sizes and properties of materials and recommending their proper use in building.

Broken bond – the use of part bricks to make good a *bonding* pattern where dimensions do not allow regularized bond patterns of full bricks.

Bullnose – *special* shaped brick with a curved surface joining two adjacent faces.

Bullseye – a circular opening in brickwork formed with a complete ring of *voussoirs*.

Calcium silicate brick – a brick made from lime and sand (*sandlime*) and possibly with the addition of crushed flint (*flintlime*), and *autoclaved*.

Cant – *special* shaped brick with a splayed surface joining two adjacent faces.

Capping – construction or component at the top of a wall or *parapet* not providing a weathered overhang (see also *coping*).

Cavity tray – see *DPC tray*.

Cavity wall – wall of two *leaves* effectively tied together with *wall ties* with a space between them, usually at least 50 mm wide.

Cement – see *Portland cement* and *masonry cement*.

Cill – see *Sill*.

Clamp – a large stack of moulded, dried *clay bricks* containing crushed fuel, which is then fired.

Classical – a style of architecture inspired by and developed from the buildings of ancient Greece and Rome.

Clay brick – a brick made from clay formed in a moist condition, dried and fired in a *kiln* or *clamp* to produce a hard semi-vitreous unit.

Closers – *bonding bricks* which expose a half *header* in the surface of a wall.

Club hammer – heavy hammer used for striking a *bolster* when cutting bricks.

Collar joint – a continuous vertical joint within a wall and parallel to its face.

Common brick – a brick for general purpose applications where appearance is not of significance.

Compressive strength – the average value of the crushing strengths of a sample of bricks tested to assess load bearing capability.

Concrete – a mixture of *sand*, gravel, *cement* and water that sets and hardens.

Concrete brick – a brick made from crushed rock aggregate bound with *Portland cement*.

Consistence – the degree of firmness with which the particles of a mortar cohere.

Coordinating size – size of a coordinating space allocated to a brick, including allowance for joints and tolerances (see also *work size* and *actual size*).

Coping – construction, or component, at the top of a wall that is weathered and grooved, and overhangs the wall surface below to throw water clear and provide protection against saturation (see also *capping*).

Corbel – a feature, or one or more projecting *courses* to form a support.

Course – a row of bricks or other *masonry units* bedded in *mortar*, generally horizontally.

Coarse stuff – a mixture of *sand* and *lime* to which *cement* and water are added to make *mortar*.

Cross-joint – vertical mortar joint at right angles to the face of the wall (sometimes incorrectly called a *perp*).

Dentilation – a decorative course or courses in which alternate *headers* project to give a toothed appearance.

Diaper – decorative pattern of diagonal intersections or diamond shapes produced by contrasting coloured bricks in a *bond* arrangement.

Dogleg – *special* shaped angle brick.

Dog toothing – a decorative course or courses in which two faces are set at 45° to the face of the brickwork, either recessed or projecting.

DPC – damp-proof course – a layer or strip of impervious material placed in a joint of a wall, chimney or similar construction to prevent the passage of water.

DPC tray – a wide *DPC* bedded in the outer and inner leaves of a cavity wall, and stepping down so that it diverts water in the cavity through *weepholes* in the outer leaf.

DPM – damp-proof membrane – a layer or sheet of impervious material within or below a floor, or vertically within or on a wall, to prevent the passage of moisture.

Dressings – masonry used to form decorative features at window and door openings and quoins.

Durability – the ability of materials to withstand the potentially destructive action of freezing conditions and chemical reactions when in a saturated state.

Eaves – lower edge of a pitched roof, or edge of a flat roof.

Efflorescence – a white powdery deposit on the face of brickwork due to the drying out of soluble salts washed from the bricks following excessive wetting.

Engineering brick – a type of clay brick traditionally used for civil engineering work for which characteristics of great strength and density are considered beneficial. They are defined by compliance with minimum *compressive strength* and maximum *water absorption* values.

European Standard – European Community standards defining the sizes and properties of materials and their proper use in building.

Extruded wire-cut bricks – bricks formed by forcing stiff moist clay, under pressure, through a die and cutting the extruded shape into individual bricks with taut wires.

Face work – brickwork built neatly and evenly without applied finish.

Facing brick – a brick for use in the exposed surface of brickwork where a consistently acceptable appearance is required.

Flared *header* – a brick with a dark coloured header due to a reduction in the amount of air or higher temperature in the vicinity during firing.

Flashing – waterproof sheet material, usually lead, dressed to prevent entry of rain water at an abutment between roof and brickwork.

Fletton bricks – *semi-dry pressed bricks* made from Lower Oxford clay. Originally made in Fletton, near Peterborough, and subsequently widely used throughout the UK.

Flintlime brick – see *calcium silicate brick*.

Foundation – a sub-structure to bear on supporting sub-soil.

Frog – generally one, but sometimes two indentations in one or both bed faces of some types of moulded or *pressed bricks*.

Frost damage – the destructive action of freezing water in saturated materials.

Gable – portion of a wall above eaves level that encloses the end of a pitched roof.

Gatehouse – a two, or more, storey entry, originally fortified, to a large dwelling, castle or town. Sometimes called a tower.

Gauge – the vertical setting-out of brick *courses*.

Gauge rod – batten marked at intervals with the appropriate *gauge*.

Gauged brickwork – brickwork built to fine tolerances, sometimes using bricks

which have been ground or otherwise produced to accurate sizes. Usually built with fine joints of 3 mm thickness or less. (see *Rubbers*).

Gauged arch – an *arch* formed of wedge-shaped bricks jointed with non-tapered mortar joints.

Gauge boxes – boxes of specific volumes to measure accurately the proportions of *cement, lime* and *sand* when preparing *mortar.*

Gault – a clay associated with chalk deposits of East Anglia. Generally bricks made with gault clay are cream or yellow but they may be light red.

Glazed bricks – clay bricks with a high gloss surface produced by coating with a clay slip or ceramic glaze before firing.

Gothic – the name usually given to the medieval style of architecture featuring the pointed form of arch.

Hand-made bricks – see *soft-mud bricks.*

Hanseatic League – A federal union of independent towns whose merchants virtually monopolized trade in the Baltic and the North Sea. At the height of its powers in the mid 14th century.

Header – the end face of a standard brick.

Joint profile – the shape of a mortar joint finish.

Jointer – a tool used to form a mortar *joint profile.*

Jointing – forming the finished surface profile of a mortar joint by tooling or raking as the work proceeds, without *pointing.*

Kiln – a permanent enclosure in which clay bricks are fired.

Label – a horizontal moulding over a window or door opening to deflect *rainwater run-off*; also called a dripstone.

Lateral load – force acting horizontally at right angles to the face of a wall. It may result from wind, retained earth or the action of associated structures.

Leaf – one of two parallel walls that are tied together as a *cavity wall*.

Level – the horizontality of *courses* of brickwork.

Lime (hydrated) – a fine powdered material, with no appreciable setting and hardening properties, used to improve the workability and water retention of *cement* based *mortars*.

Lime (hydraulic) – a fine powdered material which, when mixed with water, slowly sets and hardens and binds together to form a solid material. Traditionally used as a constituent of *mortar*.

Lime putty – slaked lime, sieved and mixed with water, possibly with a little fine sand, to form a white *mortar*. Traditionally used for thin joints in *gauged arches* and for *tuck pointing*.

Lime stain – also known as bleed or bloom. White insoluble calcarious deposits on the face of brickwork derived from *Portland cement* mortars which have been subjected to severe wetting during setting and hardening.

Line (1) – a string line used to guide the setting of bricks to *line* and *level*.

Line (2) – the straightness of brickwork.

Lintel – a component of reinforced concrete, steel or timber to support brickwork over an opening.

Machicolations – *corbelling* out to provide vertical holes down which to drop missiles or fluids on an enemy below.

Marl – a type of clay with a natural lime content.

Masonry – assemblage of *masonry units* laid and joined together with mortar.

Masonry units – a preformed component, intended for use in masonry construction.

Masonry cement – a pre-mixed blend of *Portland cement*, filler material and an *air entrainer* used to mix with *sand* and water to form a complete mortar.

Mortar – a mixture of *sand* and *cement* or *lime*, or all three, possibly with the inclusion of an air entrainer, that hardens after application and is used for jointing brickwork or as *render*.

Mortar, ready-mixed – factory-made mortar mix, usually of hydrated lime and sand, 'coarse stuff', to which water is added on site.

Mortar, ready-to-use retarded – factory-made mortar containing a retarder to delay the set for sufficient time to allow for delivery and use on site.

Movement joint – a continuous horizontal or vertical joint in brickwork designed to accommodate movement.

Mullions – vertical members dividing a window.

Oriel – a window *corbelled* out from the face of a wall.

Packs of bricks – bundles of bricks secured by bands or straps to facilitate mechanical handling.

Padstone – A masonry unit, incorporated in a structure, that distributes or spreads a concentrated load.

Palladian – an architectural style based on the published work of the Italian architect Andrea Palladio, 1518–80, following his study of the buildings of ancient Rome and advocated in England by Inigo Jones, 1573–1652.

Parapet wall – upper part of a wall that bounds a roof, balcony, terrace or bridge.

Perforated bricks – *extruded wire-cut* bricks with holes through from *bed face* to *bed face*.

Perpends (perps) – notional vertical lines controlling the verticality of *cross-joints* appearing in the face of a wall.

Pier – local thickening of a wall to improve its stiffness and resistance to *lateral loads*.

Pigments – powdered or liquid materials which may be added to a *mortar* mix in small quantities to modify its colour.

Plinth – visible projection or recess at the base of a wall or *pier*.

Plinth brick – *special* shaped brick chamfered to provide for reduction in thickness between a *plinth* and the rest of a wall.

Plumb – the verticality of brickwork.

Pointing – finishing a mortar joint by raking out the *jointing* mortar before it has set hard, typically to a depth of 15 mm, and filling with additional mortar, and tooling or otherwise working it.

Polychromatic brickwork – decorative patterned brickwork which features bricks of different colours.

Portland cement – a fine powdered material which, when mixed with water, sets and binds together to form a hard, solid material. It is used as a component of *mortar* and *concrete*.

Pressed bricks – bricks formed by pressing moist clay into shape by a hydraulic press.

Quoins – Angle or corner of a masonry wall at a *return*.

Racking back – temporarily finishing each brickwork *course* in its length short of the course below so as to produce a stepped diagonal line to be joined with later work.

Radial brick – *special* shaped brick of curved form for use in brickwork curved on plan.

Rainwater run-off – rainwater that flows over a surface before being shed.

Reference panel – a panel of brickwork built at the commencement of a contract

to set standards of appearance and workmanship.

Reinforced brickwork – brickwork incorporating steel wire or rods to enhance its resistance to *lateral load*.

Render – *mortar* applied to a wall surface as a finish.

Repointing – the raking out of old *mortar* and replacing it with new (see also *pointing*).

Retaining wall – a wall that provides lateral support to higher ground at a change of level.

Returns – the part of a wall immediately following a change of direction and the areas of walling at piers or recesses which are at right angles to the general face of the wall.

Reveal – the area of walling at the side of an opening which is at right angles to the general face of the wall.

Reverse bond – *bonding* in which asymmetry of pattern is accepted across the width of an opening or at *quoins* of a wall in order to avoid *broken bond* in the work.

Rough arch – an *arch* of standard bricks jointed with tapered mortar joints.

Rubble walls – large irregular stones or flints built with *mortar* but not in regular *courses*.

Rubbers – soft clay bricks, made to be easily ground (rubbed) to accurate sizes for use in *gauged brickwork*.

Sample panel – a panel of brickwork which may be built to compare material and workmanship with those of a *reference panel*.

Sand – a fine aggregate which forms the bulk of *mortar*.

Sandlime brick – see *calcium silicate brick*.

Sealant – a stiff fluid material which sets but does not harden. Used to exclude wind-driven rain from *movement joints* and around door and window frames.

Semi-dry pressed bricks – *clay bricks* formed by pressing semi-dry or damp, ground granular clay into shape by hydraulic press.

Shale – a type of clay, often associated with coal measures.

Sill – the lower horizontal edge of an opening.

Size – see *coordinating size, work size* and *actual size.*

Skewback – brickwork, or *special* shaped block, which provides an inclined surface from which an *arch* springs.

Soft-mud bricks – bricks moulded to shape from clay in a moist, mud-like state. Often hand-made.

Soldier – a brick laid vertically on end with the *stretcher face* showing in the surface of the work.

Soldier course – a *course* of *soldier* bricks.

Specials – bricks of special shape or size used for the construction of particular brickwork features.

Springing – plane at the end of an *arch* which springs from a *skewback.*

Squint – *special* brick for the construction of non right angled corners (see also *angle*).

Stock bricks – *soft-mud bricks*, traditionally hand-made, but now often machine moulded.

Stop – *special* shaped brick to terminate runs of *plinth, bullnose* or *cant* bricks.

Stop end – a three-sided box-shaped shoe of *DPC* material sealed to the end of a *DPC tray* to divert the discharge of water.

Storey rod – *gauge rod* of storey height with additional marks to indicate fea-

tures such as *lintel* bearings, *sills*, floor joists, etc.

Stretcher – the longer face of a brick showing in the surface of a wall.

Stucco – smooth finished *render* usually painted.

Suction rate – the tendency of a brick or block to absorb water from the *mortar* used for its bedding and *jointing*. Dense vitrified bricks have a low suction rate, porous bricks have a higher suction rate.

Sulfate attack – the chemical reaction of soluble sulphates from the ground or certain types of bricks with a chemical constituent of *Portland cement* which results in expansion of, and physical damage to, *mortar.*

Thermoluminescence tests – a laboratory method of dating minerals and ceramics by measuring their emission of light.

Ties – see *wall ties.*

Tolerance – allowable variation between a specified dimension and an actual dimension.

Tracery – The ornamental pattern-work in brick or stone in the upper part of a *Gothic* window.

Transom(ed) – horizontal divisions or crossbars of a window.

Trowel – hand tool with a thin flat blade, usually diamond shaped, for applying *mortar.*

Tuck pointing – grooves cut into brickwork to which are applied fine lines of *lime putty* to give a regular and accurate joint pattern. See 2.2.3 h.

Units – see *Masonry units.*

Vitrified – clay or sand turned into a glass-like substance due to heating.

Voussoir – a wedge-shaped brick or stone used in a *gauged arch.*

Verge – sloping edge of a pitched roof.

Wall ties – a component, made of metal or plastic, either built into the two leaves of a *cavity wall* to link them, or used as a restraint fixing to tie cladding to a backing.

Water absorption – a measure of the density of a brick by calculating the percentage increase in the weight of a saturated brick compared with its dry weight.

Work size – the size of a brick or block specified for its manufacture. It is derived from the *coordinating size* less the allowance for mortar joints (see also *actual size*).

BRIDGWATER COLLEGE LIBRARY

References and further reading

Brick Development Association*

1988: *Flexible Paving with Clay Pavers*. Smith, R.

1991: *Brickwork – Good Site Practice*. Knight, T.

1991: *The Design of Curved Brickwork*. Hammett, M and Morton, J.

1993: *The Use of Bricks of Special Shapes*. Hammett, M.

1993: *Achieving Successful Brickwork*. Several expert authors. A loose-leaf workshop reference manual in colour providing detailed support material for bricklaying tutors.

1994: *The BDA Guide to Successful Brickwork*. Edward Arnold, London. A softback, bound, low-priced version of the above for individual tutors and students.

1995: *Rigid Paving with Clay Pavers*. Hammett, M and Smith, R.

British Standards Institution

BS 187: 1978 Calcium silicate (sandlime and flintlime) bricks.

BS 3921: 1965 Bricks and blocks of fired brickearth, clay or shale. (Imperial dimensions).

BS 3921: 1969 Bricks and blocks of fired brickearth, clay or shale. Part 2. Metric units.

BS 3921: 1985 Clay bricks.

BS 4729: 1990 Dimensions of bricks of special shapes and sizes

BS 5628: Part 3 Code of practice for the use of masonry.

BS 6073: Part 1 Specification for precast concrete masonry units.

BS 6649: 1985 Clay and calcium silicate modular bricks.

Brunskill, R W 1990: *Brick Building in Britain*. Gollancz, London.

Firman R J 1994: 'The colour of brick in historic brickwork'. *Information*[†] No. 61, February.

Hammond, M D P 1984: 'Brick kilns: an illustrated survey – II Clamps'. *Information* No. 33, May.

Hammond, M D P 1985: 'Brick kilns: an illustrated survey – III Suffolk kilns'. *Information* No. 35, February.

Kennell, R B 1993: 'English Garden-Wall Bond – More than just a bond'. *Information* No. 58, February.

Kennett, D H 1984: 'Decorative brickwork in High Street, Winslow, Buckinghamshire: A preliminary Survey.' *Information* No. 33, May.

Kennett D H 1985: 'Rat-trap Bond and Flemish Bond'. *Information* No. 34, November.

Kennett D H 1985: 'English Bond at Cardington, Bedfordshire.' *Information* No. 35, February.

Kennett, D H 1985: 'Headers as a decorative feature'. *Information*, No. 36, May.

Knight, T L and Hammett, M 1993: 'The Interaction of Design and Weathering on Masonry Constructions'. *Masonry International*[tt], summer 1993.

Lloyd, N 1983: *A History of English Brickwork*. Antique Collectors Club Ltd. First published 1925, Montgomery, London.

Lynch, G 1994: *Brickwork. History, Technology and Practice*. Donhead, London.

Muthesius S 1972: *The High Victorian Movement in Architecture, 1850–1870*. Routledge & Kegan Paul, London and Boston.

Smith, T P 1986: 'Three brick churches by Sir Giles Gilbert Scott (using English Garden-Wall Bond)'. *Information* No. 38, February.

Smith, T P 1992: 'The diaper work at Queens College, Cambridge'. *Information* No. 55, May.

Tatton-Brown, T 1986: 'Hampton Court Palace'. *Information* No. 39, May.

Wight, J A 1972: *Brick Building in England. From the Middle Ages to 1550*. John Baker, London.

Woodforde, J. 1976: *Bricks to Build a House*. Routledge & Kegan Paul.

* A priced Publications List may be obtained from the Brick Development Association Association at Woodside House, Winkfield, Windsor, Berkshire SL4 2DX Tel: 01344 885651.

[t] British Brick Society, *Information*. Details of the Society may be obtained from the Honorary Secretary, c/o The Brick Development Association at the address above.

[tt] *Masonry International* is the journal of the British Masonry Society, c/o British Ceramic Research Ltd., Queens Road, Penkhull, Stoke-on-Trent ST4 7LQ.

Index